●大学英语拓展课程"十三五"规划教材●

珠宝英语
English for Jewelry

◎主　编　陈丽兰
◎副主编　叶玉珊　苏东彦　黄姗姗
◎主　审　黄运亭

华南理工大学出版社
SOUTH CHINA UNIVERSITY OF TECHNOLOGY PRESS
·广州·

图书在版编目（CIP）数据

珠宝英语/陈丽兰主编. —广州：华南理工大学出版社，2019.2（2022.7 重印）
（大学英语拓展课程"十三五"规划教材）
ISBN 978-7-5623-5312-6

Ⅰ.①珠… Ⅱ.①陈… Ⅲ.①宝石–英语–高等学校–教材
②首饰–英语–高等学校–教材 Ⅳ.①TS934.3

中国版本图书馆 CIP 数据核字（2017）第 287177 号

珠宝英语

陈丽兰 主编

出 版 人：柯 宁
出版发行：华南理工大学出版社
（广州五山华南理工大学 17 号楼，邮编：510640）
http://hg.cb.scut.edu.cn E-mail：scutc13@scut.edu.cn
营销部电话：020-87113487 87111048（传真）
总 策 划：乔 丽
策划编辑：吴翠微
责任编辑：陈 蓉
印 刷 者：广州小明数码快印有限公司
开 本：787mm×1092mm 1/16 印张：11.5 字数：355 千
版 次：2019 年 2 月第 1 版 2022 年 7 月第 2 次印刷
定 价：32.00 元

版权所有 盗版必究 印装差错 负责调换

前　言

在经济全球化、科学技术一体化的时代，作为当今世界使用最为广泛的信息载体和交流工具，英语发挥着越来越重要的作用。随着我国对外交流日益频繁，国家和社会对于大学英语教学提出了更高的要求。《珠宝英语》就是为了满足珠宝专业大学英语教学需要而编写的。

《珠宝英语》是针对珠宝专业大二、大三学生编写的英语教材，旨在帮助学生通过阅读与主修专业相关的篇章提高综合运用英语的能力。本书由10个单元构成，分别从珠宝简介、宝石特征和分类、知名珠宝及其历史、珠宝鉴定、钻石、宝石（红宝石、蓝宝石、绿宝石）、多晶宝石、珍珠、珠宝设计、珠宝商业方面对珠宝及其相关行业进行比较系统、全面的介绍。每个单元分为四个部分：引入（Starting Out）、对话（Communicative Activities）、阅读和练习（Read and Explore）、拓展阅读（Extended Learning）。练习题型多样，既有对课文知识点的复习检测，又有词汇、语法、翻译、听写等语言基本功的训练，还有小组活动项目。在每单元的最后一个部分，还附有该单元的词汇和短语。

每单元设有与该单元主题相关的引入和对话。这一部分除了提供基本句型和话题外，还为学生自己做对话练习提供了相关知识储备，即学即用，可以明显地强化学习效果。每单元的三篇阅读文章中，第一篇作为精读文章，配有大量与文章内容相关的词汇、语法、翻译、听写等练习，以实现通过阅读篇章达到提高综合运用英语能力的目的。另外两篇为拓展阅读，在介绍珠宝专业知识的同时帮助学生复习、巩固所掌握的英语知识。每单元还设计了操练项目，通过合作学习，学生可以把所学到的知识运用到实践之中。各单元后附设相关词汇、短语，方便学生参考和复习。每单元的练习都是基于该单元精读的篇章内容设计的，通过练习，学生能更好地掌握所学知识，对阅读篇章有更深刻的认识，达到复习巩固之目的。同时，以专业知识为载体，全面培养学生听、说、读、写、译等方面的语言基本技能也是本书的特色之一。语言学习需要一个过程，只有不断操练，学生才能将所学知识内化和掌握。

本教材选材新颖、素材翔实、图文并茂，可作为珠宝专业学生的英语教材，也可为珠宝设计人员以及爱好珠宝英语的人士提供参考。

鉴于编者水平有限，疏漏与不足在所难免，恳请广大读者和同行、专家批评指正。

编　者
2019年1月

CONTENTS

Unit 1　Introduction to Gemology ·· 1
　　　Starting Out ··· 1
　　　Part Ⅰ　Communicative Activities ································· 2
　　　Part Ⅱ　Read and Explore ··· 4
　　　Part Ⅲ　Extended Learning ······································· 8

Unit 2　Characteristics and Classification of Gems ··············· 14
　　　Starting Out ··· 14
　　　Part Ⅰ　Communicative Activities ······························· 15
　　　Part Ⅱ　Read and Explore ······································· 16
　　　Part Ⅲ　Extended Learning ····································· 21

Unit 3　Famous Jewelry and Their History ························· 33
　　　Starting Out ··· 33
　　　Part Ⅰ　Communicative Activities ······························· 34
　　　Part Ⅱ　Read and Explore ······································· 36
　　　Part Ⅲ　Extended Learning ····································· 39

Unit 4　Jewelry Appraisal ··· 44
　　　Starting Out ··· 44
　　　Part Ⅰ　Communicative Activities ······························· 45
　　　Part Ⅱ　Read and Explore ······································· 47
　　　Part Ⅲ　Extended Learning ····································· 51

Unit 5　Diamonds ··· 56
　　　Starting Out ··· 56
　　　Part Ⅰ　Communicative Activities ······························· 57
　　　Part Ⅱ　Read and Explore ······································· 59
　　　Part Ⅲ　Extended Learning ····································· 63

Unit 6　Ruby, Sapphire and Beryl (Emerald) ······················ 69
　　　Starting Out ··· 69
　　　Part Ⅰ　Communicative Activities ······························· 71

Part II	Read and Explore	72
Part III	Extended Learning	76

Unit 7 Polycrystalline Gemstones 80
Starting Out 80
Part I Communicative Activities 82
Part II Read and Explore 84
Part III Extended Learning 89

Unit 8 Pearl 96
Starting Out 96
Part I Communicative Activities 97
Part II Read and Explore 99
Part III Extended Learning 103

Unit 9 Gemstone Design 108
Starting Out 108
Part I Communicative Activities 109
Part II Read and Explore 111
Part III Extended Learning 115

Unit 10 Jewelry Commerce 121
Starting Out 121
Part I Communicative Activities 122
Part II Read and Explore 124
Part III Extended Learning 128

Key to the Exercises and Translation 135

References 175

Unit 1

Introduction to Gemology

Starting Out

☞ **Match Words with Pictures**

Match the words with the corresponding pictures.

> diamond   pearl   crystal   jadeite   ruby   gold

1. _____   2. _____   3. _____

4. _____   5. _____   6. _____

☞ **Check Your Knowledge**

Fill in the blanks with corresponding Chinese meanings.

1. plated gold _____
2. platinum _____
3. copper _____
4. zircon _____
5. alloy _____
6. silver _____
7. cat's eye _____
8. acrylic _____
9. agate _____
10. tourmaline _____
11. barrette _____
12. pendant _____
13. brooch _____
14. bangle _____
15. anklet _____
16. bracelet _____
17. necklace _____
18. set jewelry _____
19. buckle _____
20. scarves-tip _____

Part I   Communicative Activities

Task 1   Conversations

Ⅰ. **Read the following conversation and underline some useful sentences for buying and selling jewelry.**

(C: customer; S: jewelry salesman)

C: I'd like to buy a jadeite ring. Can you give me some suggestions?

S: Yes, it's my pleasure. What's your size?

C: I have no idea.

S: Would you mind my measuring the size for you?

C: No, I don't mind.

S: Your size is 13. What do you think of this ring?

C: I don't think the style suits me. I like the ring on the catalogue.

S: I'm sorry, but it's just for display and we don't sell it.

C: Can I try this ring on?

S: You have a good taste. It's the latest fashion, very popular.

C: Thank you. I think I'll enjoy wearing it. Check, please.

S: Generally you can't return or exchange the jewelry products. So you'd better double-check it before you take it.

C: There is no problem. How about your after-sale service?

S: We will clean and wash it for you for life. Repairing will be free of charge in one year, and then there'll be a charge for repairing cost. If you find any quality problem when you wear it, please contact us as soon as possible. We'll treat with it in time. And you can enjoy these after-sale services at our branches in China. Would you like to wrap it for a present?

C: Yes, please.

S: Here is your ring. This is the guarantee certificate, and this is the maintenance instruction, and this is the identification certificate.

C: What does Jade A on the certificate mean?

S: Jade is also known as Jadeite. The different types of Jade A, B and C represent the different manufacture processes that they have gone through. Jade A is the natural jadeite, without any treatment.

C: Thank you!

Ⅱ. Look at the following conversation. Decide where the following sentences go and then act out the conversation in pairs.

> a. They include Carat, Cut, Color and Clarity.
> b. There is a 10 points diamond solitaire for each earring, and the total weight of diamond is 20 points.
> c. What's the price range you have in mind?
> d. So we can guarantee the quality of all our diamonds.
> e. Can you recommend some to me?
> f. "Point" is the measuring unit of diamond.

(S: jewelry salesman; C: customer)

S: Is there anything I can do for you?

C: Yes. I'd like to have a pair of diamond earrings. 1. _____

S: Of course. 2. _____

C: About 2,000 *yuan*.

S: How about this pair of earrings?

C: How big is the diamond on the earrings?

S: 3. _____ Would you like to try it on?

C: It looks good on me. What does the point mean?

S: 4. _____ The international standard of diamond's weight measuring unit is "carat". It is usually abbreviated as "ct". We also take the "point" as the measuring unit for the diamond which weighs less than 1 carat. 1 carat is divided into 100 points.

C: Oh, I see. By the way, does your diamond come from South Africa?

S: Of course, some of our diamonds come from South Africa. But according to the international convention, diamonds are graded by 4C evaluation criteria, not by their source areas, so I can't accurately tell you which one is from South Africa. Our company is licensed as one of the 84 special sight holders of DTC, the world's largest diamond supplier and has obtained a direct supply of rough stones. 5. _____

C: What are the 4C evaluation criteria for diamond?

S: 6. _____

Task 2　Role-play

Work in pairs and act out the following roles in the conversation about buying and selling jewelry.

Student A: a jewelry salesman
Student B: a customer

Useful Expressions and Sentences

1. Buying jewelry

- Can you give me some suggestions? 你能给我一些建议吗？
- Can you recommend some…to me? 你能推荐一些……给我吗？
- I don't think the…suits me. I like the…on the catalogue. 我觉得……不适合我。我比较喜欢商品目录上的……
- Thank you. I think I'll enjoy wearing it. Check, please. 谢谢，我想戴上会很好看。请结账。
- How about your after-sale service? 你们的售后服务有哪些？

2. Selling jewelry

- Is there anything I can do for you? 有什么我可以帮忙的吗？
- Would you mind my measuring the size for you? 我能帮你量量尺寸吗？
- What do you think of…? 你觉得……怎么样？
- What's the price range you have in mind? 你的预算价格是多少？
- You have a good taste. It's the latest fashion, very popular. 你的眼光真好。这是最新款，很受欢迎。
- Generally you can't return or exchange… So you'd better double-check it before you take it. 一般情况下，……不能退换。所以你结账前最好再仔细检查一下。
- This is the guarantee certificate, and this is the maintenance instruction, and this is the identification certificate. 这是质保书，这是维护说明书，这是鉴定证书。
- We can guarantee the quality of all our… 我们所有的……都有质量保证。
- We will clean and wash it for you for life. Repairing will be free of charge in one year, and then there'll be a charge for repairing cost. If you find any quality problem when you wear it, please contact us as soon as possible. We'll treat with it in time. 终身免费清洗，一年内免费维修，一年后维修收取维修费。如果穿戴时发现任何质量问题，请尽快联系我们。我们将及时处理。

Part II Read and Explore

Jewelry Fashion Tips for Women

Once you're dressed in your pants suit, jeans and blouse, or even your little black dress, you can't forget the finishing touches. Jewelry is an important part of fashion that should not be neglected, even if all you put on is a gold band on your thumb. There are a few jewelry fashion tips

to keep in mind when you are finishing off your look.

Bracelets

Any time you wear what can be considered a blouse, especially one with short sleeves or three-quarter sleeves, you should wear a bracelet on one of your arms. If you are going for a casual look, especially with jeans, try silver or gold bangles or a wide metal or acrylic slip-on bracelet. For an edgy look, such as with an asymmetrical dress, try a metal cuff bracelet (shaped like a "C"). For a formal or upscale look, stick with an elegant and simple diamond tennis bracelet or a mesh bracelet made of a precious metal. Avoid wearing bracelets on both wrists, as that can overwhelm your look.

Necklaces

The necklace that you wear will depend mostly on the neckline of your blouse or dress. If you're wearing a V-neck or scoop-neck top, go with a drop necklace with a charm on the end. Wear a precious gem, crystal or rhinestone charm, with both formal and casual wear. Choker necklaces look nice with strapless dresses that expose the chest and spaghetti strap styles.

Earrings

Details are everything. Even the smallest detail, like a diamond stud, could be the perfect accessory to complete your outfit. There are three main types of earrings that are popular for women in fashion—hoops, drop earrings and studs.

Hoop earrings should be worn with casual wear, though you can feel free to wear small hoops (the width of a finger) with formal outfits—especially when the hoops are diamond-studded or rhinestone-studded. Drop earrings are generally extravagant or artistic. They are meant to be admired and commented upon. Put on a pair of drop earrings when you have a special occasion—an event where you will surely be "seen". Most styles of drop earrings go nicely with formal gowns (you see these frequently at red carpet events), dressy pant suits and jeans, when you are wearing a head-turning blouse. Diamond or other gem studs are best for casual, everyday outfits.

Less is More

Do not overdo your jewelry when accessorizing. Depending on the styles you choose, there are situations where just one interesting piece of jewelry, like a colorful rhinestone bracelet, or a pair of chandelier earrings, will be enough for the whole outfit. One of the rare acceptable situations when you could wear a necklace, earrings, and a bracelet is when you have a simple drop diamond charm necklace, diamond stud earrings and a tennis bracelet.

Check Your Understanding

Ⅰ. **Fill in the form about the match of different looks with different types of fashion jewelry according to the text.**

look/style	bracelet	earrings	necklace
casual look			

（续上表）

look/style	bracelet	earrings	necklace
edgy look			
formal or upscale look			
a V-neck or scoop-neck top			
strapless dresses/spaghetti strap styles			

II. Answer the following questions.

1. What is the most important principle suggested by the author in wearing fashion jewelry?
2. How many pieces of jewelry does the author recommend us to wear in one look?
3. In what case are bracelet, necklace and earrings acceptable in one whole outfit?
4. What are the differences between hoop, drop and stud earrings?
5. Is wearing bracelets on both wrists preferred?

Subject Focus

1. Talk about your favorite jewelry type and try to explain the reasons.
2. Talk about your favorite dressing type and the suitable jewelry matches. Then exchange opinions with your classmates.
3. Suppose you are a fashion designer, design a dressing look as well as a piece of jewelry which matches the look. Then present the jewelry and the look to the class.

Language Focus

I. Subject-related Terms

Fill in the blanks with the words or expressions being defined.

1. _____ the line formed by the edge of a garment around the neck
2. _____ a crystalline rock that can be cut and polished for jewelry
3. _____ a small ornament that is fixed to a bracelet or necklace
4. _____ items such as belts and scarves which you wear or carry but which are not part of your main clothing
5. _____ a set of clothes
6. _____ a dress, usually a long dress, which women wear on formal occasions
7. _____ an imitation diamond made from rock crystal or glass or paste
8. _____ a glassy thermoplastic that can be cast and molded or used in coatings and adhesives
9. _____ a piece of clothing that you wear on the upper half of your body, for example, a blouse or shirt

10. _____ a transparent rock that is used to make jewelry and ornaments

II. Working with Words and Expressions

In the box below are some of the words and expressions you have learned in this text. Complete the following sentences with them. Change the form if necessary.

| neglect | keep in mind | casual | stick with | overwhelm |
| feel free to | head-turning | extravagant | overdo | go with |

1. We are not _____; restaurant meals are a luxury and designer clothes are out.
2. A casual blouse _____ nicely _____ a pair of diamond studs.
3. In last night's carpet event, her gown was quite _____ and attracted so much attention.
4. _____ look is suitable for everyday dressing.
5. As a finishing touch of a look, _____ your accessories is always a bad idea.
6. To achieve a fashionable style, you can _____ mix different elements in a whole look.
7. One important principle in wearing fashion jewelry, less is more, should be _____.
8. It is suggested to _____ a kind of similar style, which will to some extent shape your personal image.
9. Details, though small, cannot be _____.
10. Wearing too many pieces of jewelry just _____ her look.

III. Grammar Work

Observe the following sentences and pay special attention to the use of attributive clauses.

1. Jewelry is an important part of fashion *that should not be neglected*, even if all you put on is a gold band on your thumb.
2. The necklace *that you wear* will depend mostly on the neckline of your blouse or dress.
3. Choker necklaces look nice with strapless dresses *that expose the chest* and spaghetti strap styles.

Now correct the mistakes in the following sentences.

1. The book that I borrowed it from the library is well written.
2. The house stood at the place which the roads meet.
3. Which is known to all, the earth is round.
4. This is the girl who practice playing piano every day.
5. Can you think of anyone who's house is in the downtown?
6. Did you see the young man whom was chosen the president?
7. Is he the man who his grandfather was a soldier?
8. The day which I was to start arrived at last.
9. I've known the reason which she is so worried.
10. We should visit the university where my father teaches there.

IV. Translation

1. Once you're dressed in your pants suit, jeans and blouse, or even your little black dress, you can't forget the finishing touches.
2. For an edgy look, such as with an asymmetrical dress, try a metal cuff bracelet (shaped like a "C").
3. For a formal or upscale look, stick with an elegant and simple diamond tennis bracelet.
4. Most styles of drop earrings go nicely with formal gowns (you see these frequently at red carpet events), dressy pant suits and jeans, when you are wearing a head-turning blouse.
5. 如果是休闲打扮，可以试试牛仔裤和白色短袖衬衫，再搭配一只金色手镯。
6. 同时佩戴太多太复杂的珠宝首饰会破坏整体的美观。
7. 耳饰主要有三种类型：耳环、耳坠和耳钉。
8. 穿戴珠宝要谨记少即是多的原则。

Part III Extended Learning

Dictation

Listen to the audio and complete the following passage with the words or expressions you hear.

(R: reporter; C: Clinton Kelly; M: Mary Alice Stephenson)

R: At the Golden Globes last night. Did you watch? The thank-yous or they were tearful and heartfelt. But who should be thanking their stylists? Clinton Kelly, author of *Oh, No, She Didn't* and style expert Mary Alice Stephenson are here to talk about the "best" and the "not-so-much" there on the red carpet. Both look great. Hello.

C: Nice to meet you.

R: I love that title. *Oh, No, She Didn't*.

C: *Oh, No, She Didn't*. Do with a little snap.

R: I know that. Mary Alice, we did see some 1. _____, didn't we?

M: We sure did. There was a return to the 80s. I don't know if that's good or bad, Robin. There was Crystal Carrington, Ella Dynasty on the red carpet last night, Angelina Jolie being in her green Versace.

C: She looked amazing, I thought.

M: She did look 2. _____, but look at the shoulders, look what that is.

R: Shoulder pads?

C: Yeah.

M: Yes. Big shoulders, you know, very clean, really classic, she's got Swarovski 3. _____ all over that 4. _____, minimal jewelry, modern looking but a flash back, a little bit of flashback.

C: A flashback to the 80s, but also sort of to the 40s. It's very Emerald City, *Wizard of Oz*, right? She's like a queen of Emerald City there. Yeah, I thought she looks stunning.

R: And what else did you see, Clinton?

C: We saw also a lot of... er, who's our next?

M: Well, Anne Hathaway.

C: Yeah, Anne Hathaway. Oh, my god. She's like one of my top three 5. _____.

M: Both Clinton and I loved Ann.

R: Oh, there, she's also wearing 6. _____.

M: She's wearing Armani. Armani Privè, Roger Vivier bag. What we're seeing is, is this gorgeous flowing hair with these 7. _____. And that kind of modernized is that makes her look sexy, young and fresh, don't you think, Clinton?

C: Absolutely, and it's completely 8. _____ practically, so it dips all the way down to her ruche, which is really sexy. She looks super statuesque in it and really sort of settle herself up as a movie star, you know, real star.

M: And last time I talked to Rachel Zoe on the phone, her 9. _____, and Ann actually had two choices up to the last minute, one a little bit crazier over the top, and she went with this old Hollywood meets new, 10. _____, sexy, work for her.

R: And it really worked for her body.

C: Absolutely.

Read More

Passage 1

Six Celebs We Love for Their Fashion Sense

The public takes its fashion cues from celebrities. Since A-listers have access to the newest fashion collections before they are available in stores, their choices influence what their fans will want to buy when the clothes make it to mass retailers.

Kate Middleton has impressed the entire world with her classy, feminine sense of style. She manages to look demure without ever looking stuffy or older than she is. She single-handedly made wrap dresses popular again when she wore a navy blue one in her engagement photo. Her lace-accented wedding gown inspired the lace dresses that are currently popular. Her massive collection of tailored coats sparked a demand for long coats that have cinched waists.

Emma Watson emerged as a style icon while she was still a teenager. She favors the subtle designs of the Chanel house. She is also often seen in ethereal, flowing fabrics. Her casual wardrobe includes fashionable custom made scarves and short jackets.

Diane Kruger has a very unique sense of style. She is that rare woman who can wear couture creations that would make lesser women look foolish. Her porcelain beauty provides the perfect contrast for the angular silhouettes that she favors. She is always one of the brightest stars at the Cannes Film Festival, where her penchant for over-the-top fashion is encouraged and celebrated.

Zooey Deschanel is the perfect fashion icon for women who want to have fun with clothes. She does not take fashion seriously. Instead, she uses her clothes to enhance her adorable, quirky persona. Her style is easy to imitate. A-line dresses flatter nearly every body shape, and headbands instantly add a girlish effect. Zooey prefers flat shoes to heels, another aspect of her style that makes her look accessible.

Andrew Garfield may seem like an unlikely fashion icon, but he balances hipster tastes with maturity quite nicely. He wears slim-fitting casual pants with blazers and professional-looking shoes. He is an example of how a young man can ease into adulthood without sacrificing his identity.

Nina Dobrev is the beautiful star of *The Vampire Diaries*. She has made a name for herself in fashion circles by choosing risky prints and complementing her olive complexion with bright neons. She wears simple cuts that show off her slim figure. She has a good eye for styling as well; she leaves her hair in loose waves for casual dresses, and she pulls it back in tight updos to mimic the sharp lines of edgier gowns. Nina is perhaps the most exciting starlet to watch right now. Hopefully she will continue to make interesting sartorial choices for the length of her career.

Passage 2

Holiday Gift Idea: Initial Charms

Personalize your gifts this holiday season with initial jewelry! It's an easy and thoughtful way to give this season. 1928 Jewelry carries a variety of styles whether you're looking for a pair of dainty and petite 14K gold-dipped initial stud earrings, letter key chains, or crystal initial charm pendant necklaces.

The Gorgeous Heirloom Locket Necklace

If you're looking for an extra special gift to give this holiday season, this antique-style heirloom locket necklace is it! A collector's item, the necklace features a beautifully encrusted locket at the end of a 32 inches long antique gold tone rope chain. The locket is 3.5 inches high and 1.5 inches wide, featuring dark ruby red crystals surrounded by Swarovski crystals in a floral pattern. Made in the USA and comes in a sage green pouch.

Sparkle in Black with Midnight Gold

Looking for something edgy this fall? We know you'll all say yes when you see our glittering Signature Midnight Gold Collection! Jet and gold soldered chains were designed to create a unique set of earrings, necklaces and bracelets, and the result is beautiful. The pieces re-

mind us of a romantic starry night, or an aerial view of skyscrapers lit up when it's dark... but we'll let you do the imagining.

The "It" Fashion Color of the Season: Grey is the New Black

This season, say hello to grey because it's the new black! Whether you like it feather light or dark like charcoal, that's what we love about the color—the flexibility of its monochromatic color scheme.

Jewelry: What Metal Color Goes with Grey?

Grey goes perfectly well with silver jewelry and accessories, but when gold has a grey/silver or mixed-metal element added in, it gives a unique and luxurious accent to your look. Case in point is our stunning Diana Collection featured with grey and gold tone accents. Try it out and you'll know exactly what we're talking about...

Group Project

1. Choose one of your favorite jewelry and write a passage about it, giving as many details as possible.
2. Design a whole look for the piece of jewelry mentioned in task one and present it to the class.
3. Study the current fashion trend of jewelry design and match, then write a report about it.

Words and Expressions

celebrity /sə'lebrəti/	n.	名人
collection /kə'lekʃən/	n.	（时装、珠宝、设计等）系列
available /ə'veɪləbəl/	adj.	可获得的
feminine /'femənɪn/	adj.	女性气质的
demure /dɪ'mjʊə/	adj.	端庄的，娴静的
stuffy /'stʌfi/	adj.	古板的，老气的
inspire /ɪn'spaɪə/	vt.	激发，使产生灵感
spark /spɑːk/	vt.	鼓舞，产生火花
emerge /ɪ'mɜːdʒ/	vi.	出现，浮现
fabric /'fæbrɪk/	n.	布料
wardrobe /'wɔːdrəʊb/	n.	衣柜
porcelain /'pɔːslɪn/	n.	瓷，瓷器
angular /'æŋgjələ/	adj.	有棱角的
silhouette /ˌsɪluˈet/	n.	轮廓
enhance /ɪn'hɑːns/	vt.	促进，增强
adorable /ə'dɔːrəbəl/	adj.	可爱的
quirky /'kwɜːki/	adj.	古怪的
persona /pə'səʊnə/	n.	表面形象
imitate /'ɪmɪteɪt/	vt.	模仿

flatter /ˈflætə/	vt.	凸显
headband /ˈhedbænd/	n.	头巾，束发带
accessible /əkˈsesəbəl/	adj.	可接近的，可获得的，平易近人的
hipster /ˈhɪpstə/	n.	赶时髦的人
maturity /məˈtʃʊərəti/	n.	成熟
blazer /ˈbleɪzə/	n.	运动夹克衫
complement /ˈkɒmpləmənt/	vt.	补充，衬托
olive /ˈɒlɪv/	adj.	橄榄色的
complexion /kəmˈplekʃən/	n.	肤色
figure /ˈfɪɡə(r)/	n.	身材，身型
slim /slɪm/	adj.	苗条的，修长的
starlet /ˈstɑːlət/	n.	渴望成名的年轻女演员
updo /ˈʌpduː/	n.	高髻（头发向上梳在头顶上盘成髻）
personalize /ˈpɜːsənəlaɪz/	vt.	使个性化
mimic /ˈmɪmɪk/	vt.	模仿，效仿
thoughtful /ˈθɔːtfəl/	adj.	考虑周到的，体贴的
dainty /ˈdeɪnti/	adj.	讲究的，秀丽的
petite /pəˈtiːt/	adj.	娇小的
pendant /ˈpendənt/	n.	坠饰，挂件
antique /ænˈtiːk/	adj.	古老的，古董的
heirloom /ˈeəluːm/	n.	老物件，传家宝
encrusted /ɪnˈkrʌstɪd/	adj.	包有外壳的
locket /ˈlɒkɪt/	n.	盒式吊坠
floral /ˈflɔːrəl/	adj.	用花制作的，饰以花卉图案的
pattern /ˈpætən/	n.	图案，样式
pouch /paʊtʃ/	n.	小袋子
edgy /ˈedʒi/	adj.	前卫的
glittering /ˈɡlɪtərɪŋ/	adj.	闪闪发亮的
solder /ˈsəʊldə(r)/	vt.	焊接，连接在一起
aerial /ˈeəriəl/	adj.	空中的
skyscraper /ˈskaɪˌskreɪpə(r)/	n.	摩天大楼
charcoal /ˈtʃɑːkəʊl/	n.	木炭
flexibility /ˌfleksəˈbɪləti/	n.	灵活性
monochromatic /ˌmɒnəkrəʊˈmætɪk/	adj.	单色的
luxurious /lʌɡˈʒʊəriəs/	adj.	奢侈的
have access to		获得
mass retailer		大规模零售商
single-handedly		独自地，单独地
wrap dress		裹身裙

lace-accented	蕾丝点缀的
tailored coat	定制外套
cinched waist	束腰
style icon	时尚偶像
A-line dress	A字裙
fashion circle	时尚圈

Unit 2

Characteristics and Classification of Gems

Starting Out

☞ **Match Words with Pictures**

Match the words with the corresponding pictures.

> diamond   pearl   quartz   turquoise   amber   chrysoberyl
> emerald   jadeite   coral   peridot   ruby   opal

1. _____ 2. _____ 3. _____ 4. _____

5. _____ 6. _____ 7. _____ 8. _____

9. _____ 10. _____ 11. _____ 12. _____

☞ **Check Your Knowledge**

Fill in the blanks with corresponding Chinese meanings.

1. precious gemstone _____
2. semi-precious gemstone _____
3. chemical composition _____
4. crystal structure _____
5. crystal system _____
6. amorphous structure _____
7. optical phenomenon _____
8. organic gems _____
9. optical properties _____
10. color distribution _____
11. transparency _____
12. synthetic gems _____

Unit 2 Characteristics and Classification of Gems

Part I Communicative Activities

Recommendations and Compliments

Task 1 Conversations

Ⅰ. Read the following conversation and underline some useful sentences for buying and selling jewelry.

(S: salesgirl; C: customer)

S: Good morning. What can I do for you?
C: I want to buy a pendant to go with my necklace.
S: There are many types of pendants and the prices vary from hundreds of dollars to thousands of dollars.
C: I need something elegant but inexpensive.
S: What's your budget, madam?
C: No more than two hundred dollars.
S: OK. May I recommend this peridot pendant? It's of the latest design.
C: Oh, it looks exquisite.
S: Would you like to try it on?
C: Certainly. I'll try it.
S: Wow, it looks good on you.
C: Thank you. I'll take one.

Ⅱ. Look at the following conversation. Decide where the following sentences go and then act out the conversation in pairs.

> a. What would you recommend?   b. Wow, they look amazing on you.
> c. May I help you?   d. And the craftsmanship is excellent.
> e. I think they suit you well.

(S: salesgirl; C: customer)
S: 1. _____
C: Yes, I'm looking for a pair of earrings with gems. 2. _____
S: Here is a pair of garnet earrings. They are very popular this year. 3. _____
C: They look charming. May I try them on?
S: Yes, sure. Let me help you.
C: Thank you.

S: 4. _____ And they go well with your hairstyle.

C: Yeah, not bad. Are they of high quality?

S: Of course. 5. _____

C: OK, I will take them.

Task 2 Role-play

Work in pairs and act out the following roles in the conversation about making recommendations and compliments.

Student A: a salesperson at a jewelry shop

Student B: a customer to buy some gem ornaments

Useful Expressions and Sentences

1. **Recommendations**
   - What kind of necklace do you prefer? 您喜欢哪种类型的项链？
   - What's your budget for the ring? 您买戒指的预算是多少？
   - May I recommend this pair of earrings? 我能推荐这款耳坠吗？
   - Would you like to try this one? 您愿意试戴一下这件吗？
   - Please try this bracelet on. 请试戴一下这个手镯。

2. **Compliments**
   - The necklace suits you well. 这项链很适合您。
   - It looks good on you. 您戴上它很好看。
   - It is the latest design. 这是最新潮的设计。
   - The gemstone is of the best quality. 这个宝石品质很好。
   - It's a perfect match with your dress. 这和您的裙子很配。

Part II Read and Explore

The Science of Gemstone Classification

Have you ever wondered how gemstones are classified? Well, according to the science of mineralogy, gemstone classification begins by distinguishing various gemstone groups based on their crystal structures and associated chemical composition. Gemstone groups are then subdivided into separate types and lastly, varieties. The process seems simple enough, but in actuality, most average consumers are unaware that groups, types and varieties are different gem classification levels. The science of gemstone classification was defined by former GIA graduates, Cornelius Hurlbut and Robert Kammerling, and remarkably, their system is still used by gemologists today.

Gemstone Groups

When two or more gem types have a similar chemical composition, crystal structure or physical qualities, they are defined as a group. There are 16 gemstone mineral groups, consisting of beryl, chrysoberyl, corundum, diamond, feldspar, garnet, jade, lapis lazuli, opal, peridot, quartz, spinel, topaz, tourmaline, turquoise and zircon. Interestingly, there are several gem "groups" that are actually stand-alone gem types, such as tourmaline, zircon, topaz and spinel. However, each of these types has multiple variations available, therefore justifying group classification status.

Gemstone Types

Gemstone types have varying chemical compositions, as well as crystal structures. There are approximately 130 or more different gemstone types on the market today, with new discoveries being added. Chemical compositions can range widely, from complex mixes of various compounds to simple, single chemical elements, like diamond which is composed of only carbon. Although diamond may have a simple composition, its crystal structure is quite complex. Crystal structures can also vary from simple single structures to immensely complicated clusters of microscopic crystal. Some types may even possess a non-crystalline structure such as opal, referred to as an amorphous structure. Inorganic gems are classified based on chemical composition, as well as crystalline structure similarities, whereas organic types are classified by chemical composition only. Examples of organic species include pearl, coral, amber and ivory.

Gemstone Varieties

Gem varieties are subdivided and branched from gem types. A perfect example would be corundum, which is a group. Blue corundum is known as sapphire and red corundum is ruby, both of which are gemstone types. Sapphires that possess asterism become a variety of sapphire—star sapphire. Varietal classifications are based on optical qualities including color, optical phenomena, color distribution and transparency.

Color: Color is produced through the absorption and transmission of light at certain frequencies. Colors are described using a combination of varied hues, tones and intensity levels. Some examples of gem varieties classified by color include citrine, the yellow-gold variety of quartz; and amethyst, which is the violet variety of quartz.

Optical Phenomena: The most common of optical phenomena is iridescence. Iridescence includes traits such as orient, play of color and labradorescence (拉长石晕彩). Fire agate is an example of a gem variety classified by optical phenomena. Fire agate is simply brown agate with iridescence. As a result of the reflection of light from thin layers of limonite, red, gold, green and blue-violet colors are displayed. Some gem varieties are distinguished by chatoyancy (the cat's eye effect), such as cat's eye tourmaline or cat's eye quartz. Other types of optical phenomena include adularescence (光彩效应), aventurescence (砂金效应), asterism (the star effect) and color change.

Color Distribution: Color distribution refers to the distribution of solid colors and patterns on a gemstone. Some gemstones distribute colors and patterns so unique that a separate varietal name is needed in order to properly classify them. One example is onyx, which is essentially agate formed with parallel rather than curved band patterns.

Transparency: In the gem trade, transparency refers to a gem's ability to transmit light, and gems are often classified by their transparency. Gemstones can be transparent, translucent or opaque. Colorless quartz is an example of a transparent variety, whereas agate and moonstone are typically translucent. Both jasper and tiger's eye are considered to be opaque, because no light is passed through these varieties.

Organics & Synthetics

Organic gems are naturally occurring gems that are the result of biological processes. This is a small group of gems including jet, coral, ivory, pearl and amber.

Synthetic gems were created in a laboratory setting. Synthetic gems have "natural" composition, but they were not naturally occurring. When compared to their naturally occurring counterparts, synthetic gemstones have identical chemical composition and crystal structure, including specific gravity and other various optical qualities. It should be noted however, that not all lab created gems are considered to be synthetic. This is because some lab-grown gems utilize unnatural ingredients, such as lab-grown opal, which is composed of silica (70% ~ 80%) and bonding agents (20% ~ 30%).

Synthetics are very often confused with "simulated" gems, but simulated gems are simply imitations possessing only optical similarities. A diamond simulant known as cubic zirconia (立方氧化锆) is a perfect example of a simulated gem; it appears to be the same as natural diamond, but the chemical and crystal structures are worlds apart.

Check Your Understanding

I. Fill in the blanks according to the text.

1. Three levels of gem classification: _____, _____, _____.
2. Gem groups are defined according to _____, _____ and _____.
3. Crystal structures include _____ and _____.
4. Varietal classifications are based on _____.
5. Organic gems include _____, _____, _____, _____ and _____.

II. Answer the following questions.

1. How many gemstone groups and gemstone types are there?
2. What optical qualities do gemstones have?
3. What types of optical phenomena are there?
4. What varieties of gemstones are there in terms of transparency?
5. How do synthetic gems and simulated gems differ from each other?

Unit 2 Characteristics and Classification of Gems

Subject Focus

1. Choose one of such gemstones as beryl, garnet and quartz to elaborate the three levels of gemstone classification and give a presentation to the class.
2. Search for information about another way of classifying gemstones, namely precious gemstones and semi-precious gemstones, and discuss whether it is a reasonable way of distinguishing gemstones.
3. Search for information about simulated gems and write a passage about how to distinguish simulated gems.

Language Focus

I. Subject-related Terms

Fill in the blanks with the words or expressions being defined.

1. _____ a piece of crystal (mineral), which, in cut and polished form, is used to make jewelry or other adornments
2. _____ a unique arrangement of atoms, ions or molecules in a crystalline liquid or solid
3. _____ the relative amounts of elements that constitute the gem
4. _____ any observable events that result from the interaction of light and matter
5. _____ the physical property of allowing the transmission of light through a material
6. _____ the property of certain surfaces that appear to change color as the angle of view or the angle of illumination changes
7. _____ cat's eye effect
8. _____ gems formed from biological processes, whether animal or vegetable
9. _____ one that is made in a laboratory, but shares virtually all chemical, optical, and physical characteristics of its natural mineral counterpart
10. _____ the ratio of the density of a substance to the density (mass of the same unit volume) of a reference substance

II. Working with Words and Expressions

In the box below are some of the words and expressions you have learned in this text. Complete the following sentences with them. Change the form of words if necessary.

classify	distinguish	define	available	complex
possess	perceive	transmit	be confused with	in actuality

1. The swimming pool is _____ only in summer.
2. However, _____, it is not as wonderful as people think.
3. The police asked me if I _____ a gun.

4. The twins were so much alike that it was impossible to _____ one from the other.
5. The World Cup final is being _____ live to over fifty countries.
6. Would you _____ this gemstone as precious one?
7. John is often _____ his twin brother.
8. I can't _____ any difference between these coins.
9. It was a _____ problem.
10. It is difficult to _____ what makes him so popular.

Ⅲ. **Grammar Work**

Observe the following sentences and pay special attention to the use of passive voice.

1. Gemstone groups *are* then *subdivided* into separate types and lastly, varieties.
2. The science of gemstone classification *was defined* by former GIA graduates.
3. Inorganic gems *are classified* based on chemical composition.
4. Some gem varieties *are distinguished* by chatoyancy (the cat's eye effect).
5. It should *be noted* however, that not all lab created gems are considered to be synthetic.

Now choose the correct answers to complete the following sentences.

1. It is requested that the gemstone _____ into the shape of heart.
   A. was cut   B. to cut   C. be cut   D. is cut
2. Most commonly, gems _____ minerals.
   A. are composed of   B. compose of   C. compose   D. composing
3. The gem-forming minerals are by far the largest group, with more than 130 different minerals _____ as gems.
   A. use   B. being used   C. are used   D. are using
4. There is a special class of gemstone _____ synthetic gemstones.
   A. known as   B. known for   C. is known as   D. is known for
5. My necklace wants _____. Could you help me with it?
   A. to polish   B. polish   C. polishing   D. being polished
6. I'm afraid that we have to keep you waiting for another two days, for your ring _____.
   A. is processed   B. has been processed
   C. is processing   D. is being processed
7. The earrings _____ with this device.
   A. can mend   B. can mended   C. can be mended   D. can be mend
8. Gems _____ through human intervention are called synthetic gemstones.
   A. formed   B. form   C. forms   D. forming
9. Diamond is the hardest naturally _____ material and is also quite tough.
   A. occurred   B. occurring   C. has been occurred   D. is occurred
10. The certificate can indicate whether or not the gemstone _____.
    A. is treated   B. has been treated   C. had been treated   D. was treated

IV. Translation

1. Gemstone classification begins by distinguishing various gemstone groups based on their crystal structures and associated chemical composition.
2. However, each of these types has multiple variations available, therefore justifying group classification status.
3. There are approximately 130 or more different gemstone types on the market today, with new discoveries being added.
4. In the gem trade, transparency refers to a gem's ability to transmit light, and gems are often classified by their transparency.
5. 宝石是一种可进行切割和打磨并用于制作珠宝或其他装饰用途的材料。
6. 钻石、绿宝石、红宝石和蓝宝石都被认为是贵重珠宝，因此也最具价值，最吸引人。
7. 红宝石是红色类别的刚玉，其他颜色的刚玉称为蓝宝石。
8. 宝石可以由矿物、有机材料和其他非有机材料构成。

Part Ⅲ   Extended Learning

Dictation

Listen to the audio and complete the following passage with the words or expressions you hear.

Different Wedding Band Styles

Hi. I'm Abbott Taylor from Abbott Taylor Jewelers in Tucson, Arizona, here for About. com. Now we're going to talk about wedding band. Wedding bands can be like 1. _____, a true work of art or it can be plain and simple, starting with the plainest and most simple band. A round wedding band can 2. _____, narrower, wider, white gold, yellow gold, with no ornamentation or with a lot of ornamentation.

Eternity Wedding Band Style

One 3. _____ might be a completely paved band covered with diamonds from front to back. These are called eternity bands. An eternity band is when the diamonds go all the way around and the 4. _____ continues all the way around the piece.

In this particular case, we have some 5. _____ interlaced with pierced design work and diamonds that go around the entire piece. And, you can even have two-toned band, where you have different colors of metals used in the same band. Don't feel like you have to stay with diamonds. Diamonds are not the only way to have 6. _____ or a wedding ring.

Precious Stones in a Wedding Band

Here's an unusual example of something for an engagement ring or a wedding band, with a trillion-cut sapphire in it and diamonds. Here's a wedding band with an 7. _____ as a center stone and diamonds on the side. Very different, very unusual and it sets you aside as somebody who wants something different. There are 8. _____, flat bands, round bands, half round bands, curved bands, textured bands. There's so many different ideas.

Unusual Wedding Band Styles

This is a flat band that has a square outside shape. And, so, here we have a look at an antique filigree piece. Of course it's not antique, it's 9. _____; however, it has that feeling of antiquity. Here's filigree detail work that's just not seen commonly in most wedding bands nowadays.

So, here's something a little bit more unusual a little step from the 10. _____. And this type of setting is called scatter set. The stones are actually set below the surface with a very very soft satin finish.

Tips for Picking a Wedding Band

I recommend that when you look at a piece and you see something that tickles you in the least bit of way, keep a copy of that picture. Go from piece to piece until you find a half a dozen different ideas that ring a bell with you. And, when you find those ideas you're going to narrow down your search.

Thanks for watching. For more information, visit us on the web at About. com.

 Read More

Passage 1

Birthstones and Zodiac Stones

Birthstones are of special interest to gemstone lovers and the association of specific gemstones with months of the year goes back centuries. As the famous gemologist G. F. Kunz noted in his book, *The Curious Lore of Precious Stones*, "There is no doubt that the owner of a ring or ornament set with a birthstone is impressed with the idea of possessing something more intimately associated with his or her personality than any other stone, however beautiful or costly. The idea that birthstones possess a certain indefinable, but nonetheless real significance, has long been present and still holds a spell over the minds of all who are gifted with a touch of imagination and romance."

Precious gems were first associated with signs of the zodiac. The modern birthstone list was formulated by an organization known as The Jewelers of America, in 1912.

Unit 2 Characteristics and Classification of Gems

Birthstone and Zodiac Sign Information

January Birthstone—Garnet

The zodiac stone for Aquarius (Jan 21 – Feb 18): Garnet

Garnets are attributed with the power of protecting wearers from nightmares and snakebites, and are also thought to provide guidance in the dark. Tradition connects the stone with blood. They are a popular choice as presents for second wedding anniversary.

February Birthstone—Amethyst

The zodiac stone for Pisces (Feb 19 – Mar 20): Amethyst

The amethyst is believed to induce peace, serenity and temperance. Additionally, it is thought to restrict sensual and alcoholic intoxication, which may be good or bad. Also, amethyst is believed to improve skin and hair, and prevent baldness. It is said to protect its wearer from deceit. Its purple shade is linked to royalty and makes an ideal gift for the sixth wedding anniversary.

March Birthstone—Aquamarine

The zodiac stone for Aries (Mar 21 – Apr 20): Bloodstone

Aquamarine is associated with attributes of good health, love, youthfulness and hope. Sailors believe in its protection. Generally chosen as a 19th anniversary gift, this stone is in the beryl family, the same as emerald. Bloodstone is the astrological alternative.

April Birthstone—Diamond

The zodiac stone for Taurus (Apr 21 – May 21): Sapphire

Diamonds are traditionally linked to love, and eternal endurance because of diamond's superior hardness. The name diamond is derived from the Greek "adamas", which means invincible. Traditionally diamond is accepted as the 10th and 60th wedding anniversary presents. Sapphire is the astrological stone for Taurus.

May Birthstone—Green Emerald

The zodiac stone for Gemini (May 22 – Jun 21): Agate

Emeralds are said to cure people of bad health, healing a variety of illnesses and ensuring good health for wearers. Also, it is believed that the wearer of emerald will be able to see into the future. The astrological alternative is agate.

June Birthstone—Pearl, Moonstone

The zodiac stone for Cancer (Jun 22 – Jul 22): Emerald

The romantic pearl is associated with chastity and modesty. Stable marriages are said to be the result of pearls. The astrological stone is emerald.

July Birthstone—Ruby

The zodiac stone for Leo (Jul 23 – Aug 23): Onyx

Rubies are believed to bring harmony of life to the wearer, so one can expect a peaceful life when wearing this stone. The astrological stone is onyx.

August Birthstone—Peridot

The zodiac stone for Virgo (Aug 24 – Sep 22): Carnelian

Peridot is believed to protect the wearer from evil, particularly from the forces of darkness. It can also increase the healing power of medicinal drugs. The astrological choice is carnelian.

September Birthstone—Sapphire

The zodiac stone for Libra (Sep 23 – Oct 23): Chrysolite

Faith and purity will come to the wearer of this stone. Common belief says the brighter the glitter of the stone, the stronger the positive virtues, while giving the lucky person useful foresight into the future. The astrological stone is chrysolite.

October Birthstone—Opal, Tourmaline

The zodiac stone for Scorpio (Oct 24 – Nov 22): Beryl

Opals are said to be stones of hope and bringers of the virtues of innocence and purity. Opals stimulate healing and increase friendship and healthy emotions. Beryl is the astrological stone.

November Birthstone—Yellow Topaz, Citrine

The zodiac stone for Sagittarius (Nov 23 – Dec 21): Citrine

Cool topaz is said to hold the power of healing and preserving one's sanity. The stone is associated with life and strength of body and mind. It was once thought that the wearer would become invisible when danger was near. The astrological stone is citrine.

December Birthstone—Turquoise, Blue Topaz, Tanzanite, Blue Zircon

The zodiac stone for Capricornus (Dec 22 – Jan 20): Ruby

Turquoise is believed to bring good fortune, good luck and a happy life, things that everyone would like to have. Ruby is the astrological stone.

Passage 2

Lively Green Peridot for Sale

2.5ct Lively Green Peridot   9 mm × 8 mm

Type:	Peridot
Item ID:	334417
Content:	1 gem
Weight:	2.5 ct
Size:	8.9 mm × 7.87 mm × 4.88 mm
Shape:	Cushion Checkerboard
Clarity:	VS – SI
Treatment:	None
Origin:	China
Price:	USD 90.00

Guaranteed 100% Natural Gem

Photos above are of Actual Item for Sale.

Description

This natural Lively Green Peridot originates from China and its weight is about 2.5 carats. The measurements are 8.9 mm × 7.87 mm × 4.88 mm (length × width × depth). The shape/cut-style of this Peridot is Cushion Checkerboard. This 2.5 ct Peridot gem is available to order and can be shipped anywhere in the world. Gemstone Certification can also be provided as an optional service.

Shipping

- FedEx or DHL for USD 20.00 only / Delivery 3 – 4 days / Fully insured.
- Registered Mail for USD 6.99 / Delivery 7 – 14 days / Insured for USD 200.00.

Payment Options

- Credit Card
- PayPal
- Western Union
- Bank transfer

Guarantee/Certification

- All gemstones are natural.
- All photos and videos are of the actual gem for sale (No Stock Photos).
- We offer certification from AIGS for USD 30.00 (5 – 10 additional days).
- We also offer certification from the BGL lab for USD 15.00 only (3 – 4 days).

Size and Weight

- Gems are always measured in Millimeter (mm).
- Dimensions are given as length × width × depth, except for round stones which are diameter × depth.
- **Select gems by size, not by weight.** Gem varieties vary in density, so carat weight is not a good indication of size.
- Note: 1 ct = 0.2 g

Additional Quantities

- Every item is unique, so you can only order 1 of each item.
- However, we often have similar or matching items. If so, we display them on the same gem detail page.

Clarity Explained

- IF = Loupe Clean—internally flawless; free of inclusions.
- VVS = Almost Loupe Clean—very, very slight inclusions; inclusions barely seen under 10x magnification.
- VS = Eye Clean—very slight inclusions; inclusions can be seen under 10x magnification.
- SI = Slightly included—small inclusions can be seen with the naked eye.
- I1 = Included—inclusions can be seen with the naked eye.

- Transparent = A gemstone having the property of transmitting light without serious diffusion / may have rutile or other inclusions.
- Translucent = Allowing light to pass through, but not transparent.
- Opaque = Does not allow light to pass through.

Return Policy
- All gems have a 30-day inspection period, which begins on the date of delivery. Gems may be returned for a full refund during this period.
- Please email us for return authorization code when returning a gem.

About Peridot

Peridot belongs to the forsterite-fayalite（镁－铁橄榄石）mineral series which is part of the olivine（橄榄石）group. It is one of the "idiochromatic（自色）" gems, meaning its color comes from the basic chemical composition of the mineral itself, not from minor impurities, and therefore will only be found in shades of green. In fact, peridot is one of the few gemstones found in only one color. Historically important sources in Egypt have been replaced by today's main sources in Arizona and Pakistan. The Pakistani peridot in particular is very fine, and a new find in Pakistan in the mid-1990's has made peridot available to a wider market. The supply of peridot from China is also increasing and we have recently sourced top quality peridot from Vietnam.

Group Project

1. Write a short passage to introduce your birthstone, giving details including its chemical composition, crystal structure, physical qualities, value and significance.
2. Passage 2 in **Read More** is an introduction to a piece of gemstone for sale that is posted on www.gemselect.com. Suppose you own an online jewelry store. Now choose an item of gem and work out a webpage for it, accurately describing it with details including gem type, size, weight, shape, color, clarity and origin, providing sales information such as shipping, payment options, guarantee, and return policy and so on.

Words and Expressions

diamond /ˈdaɪəmənd/	n.	钻石
pearl /pɜːl/	n.	珍珠
quartz /kwɔːts/	n.	石英
turquoise /ˈtɜːkwɔɪz/	n.	绿松石
amber /ˈæmbə/	n.	琥珀
chrysoberyl /ˌkrɪsəˈberɪl/	n.	金绿宝石
emerald /ˈemərəld/	n.	翡翠
jadeite /ˈdʒeɪdaɪt/	n.	硬玉
coral /ˈkɒr(ə)l/	n.	珊瑚
peridot /ˈperɪdɒt/	n.	橄榄石
ruby /ˈruːbi/	n.	红宝石

opal /ˈəʊp(ə)l/	n.	欧泊，蛋白石
composition /ˌkɒmpəˈzɪʃn/	n.	组成，成分
crystal /ˈkrɪst(ə)l/	n.	结晶体
amorphous /əˈmɔːfəs/	adj.	非晶质的
optical /ˈɒptɪkəl/	adj.	光的，光学的
phenomenon /fəˈnɒmɪnən/	n.	现象，效应
organic /ɔːˈgænɪk/	adj.	有机的
property /ˈprɒpəti/	n.	性质，特性
distribution /ˌdɪstrɪˈbjuːʃ(ə)n/	n.	分布
transparency /trænˈspærənsi/	n.	透明度
synthetic /sɪnˈθetɪk/	adj.	合成的
beryl /ˈberəl/	n.	绿柱石
corundum /kəˈrʌndəm/	n.	刚玉
sapphire /ˈsæfaɪə/	n.	蓝宝石
garnet /ˈgɑːnɪt/	n.	石榴石
jade /dʒeɪd/	n.	玉石
lapis lazuli /ˌlæpɪsˈlæzjuli/	n.	天青石
spinel /ˈspɪn(ə)l/	n.	尖晶石
topaz /ˈtəʊpæz/	n.	托帕石，黄玉
tourmaline /ˈtʊərməlɪn/	n.	碧玺，电气石
zircon /ˈzɜːrk(ə)n/	n.	锆石
ivory /ˈaɪvəri/	n.	象牙
classification /ˌklæsɪfɪˈkeɪʃ(ə)n/	n.	分类
classify /ˈklæsɪfaɪ/	vt.	分类
mineralogy /ˌmɪnəˈrælədʒi/	n.	矿物学
distinguish /dɪˈstɪŋgwɪʃ/	v.	分辨
subdivide /sʌbdɪˈvaɪd/	v.	细分
define /dɪˈfaɪn/	v.	定义，确定
remarkably /rɪˈmɑːkəbli/	adv.	明显地，非常
gemologist /dʒemˈɒlədʒɪst/	n.	宝石学家
mineral /ˈmɪn(ə)r(ə)l/	n.	矿物
feldspar /ˈfeldspɑː/	n.	长石
variation /ˌveriˈeɪʃ(ə)n/	n.	变体
available /əˈveɪləb(ə)l/	adj.	有效的，可获取的
justify /ˈdʒʌstɪfaɪ/	v.	证明有理，对……做出辩护
status /ˈsteɪtəs/	n.	地位
approximately /əˈprɒksɪmətli/	adv.	大约
compound /ˈkɒmpaʊnd/	n.	复合物，组成部分
compose /kəmˈpəʊz/	v.	组成，构成

carbon /ˈkɑːbən/	n.	碳
complex /ˈkɒmpleks/	adj.	复杂的
immensely /ɪˈmensli/	adv.	极大地
cluster /ˈklʌstə/	n.	簇，群
microscopic /ˌmaɪkrəˈskɒpɪk/	adj.	微小的
possess /pəˈzes/	v.	拥有
refer /rɪˈfɜː/	v.	提到，针对
similarity /ˌsɪməˈlærəti/	n.	相似性
species /ˈspiːsiːz/	n.	品种
asterism /ˈæstərɪz(ə)m/	n.	星光效应
absorption /əbˈzɔːpʃ(ə)n/	n.	吸收
transmission /trænzˈmɪʃ(ə)n/	n.	传播
transmit /trænsˈmɪt/	v.	传播
frequency /ˈfriːkwənsi/	n.	频率
hue /hjuː/	n.	色彩
tone /təʊn/	n.	色调
intensity /ɪnˈtensəti/	n.	强度
citrine /ˈsɪtrɪn/	n.	黄水晶
amethyst /ˈæməθɪst/	n.	紫水晶
violet /ˈvaɪələt/	adj.	蓝紫色的
iridescence /ˌɪrɪˈdes(ə)ns/	n.	晕彩效应
orient /ˈɔːriənt/	n.	珍珠光泽
limonite /ˈlaɪmənaɪt/	n.	褐铁矿
display /dɪˈspleɪ/	v.	呈现，展现
chatoyancy /ʃəˈtɔɪənsi/	n.	猫眼效应
pattern /ˈpæt(ə)n/	n.	样式，图案
unique /juːˈniːk/	adj.	独一无二
onyx /ˈɒnɪks/	n.	缟玛瑙
essentially /ɪˈsenʃ(ə)li/	adv.	基本上，根本上
agate /ˈægət/	n.	玛瑙
parallel /ˈpærəlel/	n.	平行
transparent /trænˈspær(ə)nt/	adj.	透明的
translucent /trænˈsluːs(ə)nt/	adj.	半透明的
opaque /əʊˈpeɪk/	adj.	不透明的
moonstone /ˈmuːnstəʊn/	n.	月光石
typically /ˈtɪpɪkli/	adv.	典型地，通常
jasper /ˈdʒæspə/	n.	碧玉
occur /əˈkɜː/	v.	发生
jet /dʒet/	n.	黑玉

counterpart /ˈkaʊntəpɑːt/	n.	对应的事物
identical /aɪˈdentɪkəl/	adj.	完全一样的
utilize /ˈjuːtəlaɪz/	v.	利用
ingredient /ɪnˈɡriːdɪənt/	n.	成分
simulate /ˈsɪmjəleɪt/	v.	模仿，仿造
imitation /ɪmɪˈteɪʃən/	n.	仿制品，赝品
zodiac /ˈzəʊdiæk/	n.	星座
ornament /ˈɔːnəmənt/	n.	饰品，装饰物
intimately /ˈɪntɪmətli/	adv.	熟悉地，紧密地
personality /ˌpɜːsəˈnæləti/	n.	性格
costly /ˈkɒstli/	adj.	昂贵的
indefinable /ˌɪndɪˈfaɪnəbəl/	adj.	无法形容的
nonetheless /ˌnʌnðəˈles/	adv.	然而
spell /spel/	n.	魔力
formulate /ˈfɔːmjəleɪt/	v.	制定，规划
Aquarius /əˈkweərɪəs/	n.	水瓶座
attribute /əˈtrɪbjuːt/	v.	认为某物属于某人，把……归功于
nightmare /ˈnaɪtmeə/	n.	噩梦
anniversary /ˌænəˈvɜːsəri/	n.	周年纪念
Pisces /ˈpaɪsiːz/	n.	双鱼座
induce /ɪnˈdjuːs/	v.	引起，引诱
serenity /sɪˈrenəti/	n.	宁静
temperance /ˈtemp(ə)r(ə)ns/	n.	节制
sensual /ˈsenʃuəl/	adj.	肉欲的
baldness /ˈbɔːldnəs/	n.	脱发，光秃
deceit /dɪˈsiːt/	n.	欺骗
shade /ʃeɪd/	n.	色度，深浅
royalty /ˈrɔɪəlti/	n.	皇族，王室
aquamarine /ˌækwəməˈriːn/	n.	海蓝宝石
Aries /ˈeəriːz/	n.	白羊座
astrological /ˌæstrəˈlɒdʒɪkəl/	adj.	占星的，占星术的
alternative /ɔːlˈtɜːnətɪv/	n.	可供选择的事物
Taurus /ˈtɔːrəs/	n.	金牛座
eternal /ɪˈtɜːn(ə)l/	adj.	永恒的
endurance /ɪnˈdjʊər(ə)ns/	n.	持久
invincible /ɪnˈvɪnsəb(ə)l/	adj.	无法征服的
Gemini /ˈdʒemənaɪ/	n.	双子座
Cancer /ˈkænsə/	n.	巨蟹座

chastity /ˈtʃæstəti/		n.	贞洁
Leo /liːəʊ/		n.	狮子座
Virgo /ˈvɜːgəʊ/		n.	处女座
evil /ˈiːvəl/		n.	邪恶
Libra /ˈlaɪbrə/		n.	天秤座
Chrysolite /ˈkrɪsəlaɪt/		n.	贵橄榄石
faith /feɪθ/		n.	忠诚
purity /ˈpjʊərəti/		n.	纯洁
glitter /ˈglɪtə/		n.	光辉
virtue /ˈvɜːtʃuː/		n.	优点
foresight /ˈfɔːsaɪt/		n.	先见，预见
Scorpio /ˈskɔːpiəʊ/		n.	天蝎座
innocence /ˈɪnəsəns/		n.	天真无邪
stimulate /ˈstɪmjəleɪt/		v.	促进，激励
Sagittarius /ˌsædʒɪˈteəriəs/		n.	射手座
preserve /prɪˈzɜːv/		v.	保护，保存
sanity /ˈsænɪti/		n.	清醒
invisible /ɪnˈvɪzəbəl/		adj.	看不见的
tanzanite /ˈtænzənaɪt/		n.	坦桑黝帘石
Capricornus /ˈkæprəkɔːnəs/		n.	摩羯座
clarity /ˈklærəti/		n.	净度
treatment /ˈtriːtmənt/		n.	处理
carat /ˈkærət/		n.	克拉
measurement /ˈmeʒəmənt/		n.	测量，尺寸
ship /ʃɪp/		v.	运送，运输
optional /ˈɒpʃənəl/		adj.	可选择的
option /ˈɒpʃən/		n.	选项
insure /ɪnˈʃʊə/		v.	投保
delivery /dɪˈlɪvəri/		n.	递送，运送
certification /ˌsɜːtɪfɪˈkeɪʃən/		n.	证明，证书
dimension /daɪˈmenʃən/		n.	尺寸
diameter /daɪˈæmɪtə/		n.	直径
density /ˈdensəti/		n.	密度，比重
indication /ˌɪndɪˈkeɪʃən/		n.	迹象，表明
quantity /ˈkwɒntəti/		n.	数量
internally /ɪnˈtɜːnəli/		adv.	内部地
flawless /ˈflɔːləs/		adj.	无瑕的
inclusion /ɪnˈkluːʒ(ə)n/		n.	内含物
slight /slaɪt/		adj.	微小的

barely /ˈbeəli/	adv.	仅仅，几乎不
magnification /ˌmæɡnɪfɪˈkeɪʃ(ə)n/	n.	放大
diffusion /dɪˈfjuːʒ(ə)n/	n.	扩散
rutile /ˈruːtɪl/	n.	金红石
inspection /ɪnˈspekʃ(ə)n/	n.	检查，检阅
authorization /ˌɔːθəraɪˈzeɪʃ(ə)n/	n.	授权
code /kəʊd/	n.	编码
impurity /ɪmˈpjʊərɪti/	n.	杂质

in actuality	事实上
average consumer	普通消费者
physical qualities	物理性质
consist of	由……构成
be composed of	由……构成
refer to	指的是，参考
fire agate	火玛瑙
reflection of light	光折射
solid color	纯色
curved band	曲线
be compared to	与……相比
specific gravity	比重
bonding agents	黏合剂
be confused with	与……混淆
simulated gem	仿制宝石
cubic zirconia	立方氧化锆
there is no doubt that	……毫无疑问
be impressed with	对……印象不错
be intimately associated with	与……密切相关
be attributed with	具有
alcoholic intoxication	醉酒
be derived from	源自
originate from	来源于
Cushion Checkerboard	枕形棋盘式
registered mail	挂号信
Loupe Clean	镜下无瑕
free of	摆脱，无……的
naked eye	肉眼
full refund	全额退货
in particular	尤其，特别

Unit 3

Famous Jewelry and Their History

Starting Out

☞ **Match Words with Pictures**

Match the words with the corresponding pictures.

| pendant | necklace | bracelet |

1. _____ 2. _____ 3. _____

☞ **Check Your Knowledge**

Match the famous jewelry with each of the following great films.

Famous Jewelry	Film
The Heart of the Ocean Pendant	*The Lord of the Rings*
Ruby and Diamond Necklace	*Breakfast at Tiffany's*
The One Ring	*The Great Muppet Caper*
Audrey Hepburn's Pearls	*Pretty Woman*
The Baseball Diamond	*Titanic*
Diamond Necklace	*To Catch a Thief*

Part I Communicative Activities

Task 1 Conversations

Ⅰ. **Read the following conversation and underline some useful sentences for buying and selling jewelry.**

(W: Mrs. Wang; B: Bruce)

W: Well, we've settled the question of price, quality and quantity. Now what about the terms of payment?

B: We only accept payment by irrevocable letter of credit payable against shipping documents.

W: I see. Could you make an exception and accept D/A or D/P?

B: I'm afraid not. We insist on a letter of credit.

W: To tell you the truth, a letter of credit would increase the cost of my import. When I open a letter of credit with a bank, I have to pay a deposit. That'll tie up my money and increase my cost.

B: Consult your bank and see if they will reduce the required deposit to a minimum.

W: Still, there will be bank charges in connection with the credit. It would help me greatly if you would accept D/A or D/P. You can draw on me just as if there were a letter of credit. It makes no great difference to you, but it does to me.

B: Well, Mrs. Wang, you must be aware that an irrevocable letter of credit gives the exporter the additional protection of the banker's guarantee. We always require L/C for our exports. And the other way round, we pay by L/C for our imports.

W: To meet you half way, what do you say if 50% by L/C and the balance by D/P?

B: I'm very sorry, Mrs. Wang. But I'm afraid I can't promise you even that. As I've said, we require payment by L/C.

Ⅱ. **Look at the following conversation. Decide where the following expressions go and then act out the conversation in pairs.**

a. letters of credit and effect payment in yen.
b. we've settled everything in connection with this transaction.
c. It is convenient to make payment in pound sterling.
d. They'll do so against our sales confirmation or contract.

(B: Mr. Brown; S: sales manager)

S: Well, Mr. Brown, 1. _____ except the question of payment in yen. Now can you explain to me how to make payment in yen?

B: Many of our business friends in England, France, Switzerland, Italy and Germany are paying for our exports in Japan currency. It is quite easy to do so.

S: I know some of them are doing that. But this is new to me. I've never made payment in yen before. 2. _____ But I may have some difficulty in making payment in yen.

B: Many banks in Europe now carry accounts in yen. They are in a position to open 3. _____ Consult your banks and you'll see that they are ready to offer you this service.

S: Do you mean that I can open a letter of credit in yen with a bank in London or Bonn?

B: Sure you can. Several banks in London, such as the National Westminster Bank and Barclays Bank are in a position to open letters of credit in yen. 4. _____

S: I see.

Task 2 Role-play

Work in pairs and act out the following roles in the conversation about payment.

Student A: a buyer of jewelry

Student B: a supplier of jewelry

Useful Expressions and Sentences

- Our terms of payment are by a confirmed irrevocable letter of credit by draft at sight. 我们的支付方式是以保兑不可撤销的、凭即期汇票支付的信用证。
- Since the total amount is so big and the world monetary market is rather unstable at the moment, we cannot accept any terms of payment other than a letter of credit. 由于总金额太大，目前世界金融市场相当不稳定，我们不能接受信用证以外的任何支付方式。
- I'd like to discuss the terms of payment with you. I wonder if you would accept D/P. 我想跟你谈谈支付方式。想知道你方是否接受付款交单。
- We have instructed our bank to open an irrevocable documentary letter of credit in your favor. The amount is $1,300. 我们已指示我方银行开立以你方为抬头的不可撤销跟单信用证。金额是1300美元。
- We'd like you to accept D/P for this transaction and future ones. 我方希望贵方能接受本次交易起以付款交单方式来支付。
- We shall draw on you at 60 days sight the goods have been shipped. Please honor our draft when it falls due. 货物已装船，我们将于60日内向你方索取款项。到期时请兑付汇票。

English for Jewelry

Part II Read and Explore

The Taylor-Burton Diamond and Its History

Diamonds have no mercy... "They will show up the wearer if they can," says one character in *The Sandcastle*, an early novel by the famous British author, Iris Murdoch. Now this may be true of some women—usually wearing an outrageously large item of jewelry which imparts a degree of unwholesome vulgarity to themselves—but is it applicable to Elizabeth Taylor? Those well-publicized gifts which she received from her fifth husband, the late Richard Burton, certainly enhance her appearance and do not look out of place on her. A compatibility is established between the jewel and its wearer.

By far the best known of Richard Burton's purchases was the 69.42 carat pear-shape, later to be called the Taylor-Burton Diamond. It was cut from a rough stone weighing 240.80 carats found in the Premier Mine in 1966 and subsequently bought by Harry Winston. Harry commented at the time, "I don't think there have been half a dozen stones in the world of this quality."

After the rough piece arrived in New York, Harry Winston and his cleaver, Pastor Colon Jr. studied it for six months. Markings were made, erased and redrawn to show where the stone could be cleaved. After he had cleaved the stone, the 50-year-old cleaver said nothing—he reached across the workbench for the piece of diamond that had separated from it and looked at it through his horn-rimmed glasses for a fraction of a second before exclaiming "Beautiful!" This piece of rough weighed 78 carats was expected to yield a stone of about 24 carats, while the large piece, weighing 162 carats, was destined to produce a pear shape whose weight had originally been expected to be about 75 carats.

The stone's first owner after Harry Winston wasn't actually Elizabeth Taylor. In 1967, Winston sold the pear shape to the sister of the American ambassador in London during the Richard Nixon administration. Two years later, the diamond was put up for auction. Elizabeth Taylor was one name among them and she did indeed have a preview of the diamond.

The auctioneer began the bidding by asking if anyone would offer $200,000, at which the crowded room erupted with a simultaneous "Yes". Bidding began to climb, and with nine bidders active, rushed to $500,000. At $500,000 the individual bids increased in $10,000 increments. At $650,000 only two bidders remained. When the bidding reached $1,000,000, Al Yugler of Frank Pollak, who was representing Richard Burton, dropped out. Pandemonium broke out when the hammer fell and everyone in the room stood up. The winner was Robert Kenmore, the Chairman of the Board of Kenmore Corporation, the owners of Cartier Inc., who paid the record price of $1,050,000 for the gem, which he promptly named the "Cartier". The previous record for a jewel had been $305,000 for a diamond necklace from the Rovensky estate in 1957.

But Burton was not finished yet and was determined to acquire the diamond. So, speaking from a pay-phone of a well-known hotel in southern England, he spoke to Mr. Kenmore's agent. In the end, Robert Kenmore agreed to sell it, but on the condition that Cartier was able to display it, by now named the Taylor-Burton, in New York and Chicago.

Shortly afterwards on November 12th, Miss Taylor wore the Taylor-Burton in public for the first time when she attended Princess Grace's 40th birthday party in Monaco. It was flown from New York to Nice, Italy in the company of two armed guards hired by Burton and Cartier. In 1978, following her divorce from Richard Burton, Miss Taylor announced that she was putting the diamond up for sale and was planning to use part of the proceeds to build a hospital in Botswana. In June of 1979, Henry Lambert, the New York jeweler, stated that he had bought the Taylor-Burton Diamond for $5,000,000.

By December he had sold the stone to its present owner, Robert Mouawad. Soon after, Mr. Mouawad had the stone slightly recut and it now weighs 68.09 carats.

Check Your Understanding

I. **Put the names of the owners in time order according to the text.**

The Owners of The Taylor-Burton Diamond
Order: 1. _____   2. _____   3. _____   4. _____
       5. _____   6. _____   7. _____

A. Elizabeth Taylor   B. Richard Burton   C. Harry Winston   D. Henry Lambert
E. The sister of the American ambassador   F. Robert Kenmore   G. Robert Mouawad

II. **Answer the following questions.**

1. According to the text, does the Taylor-Burton Diamond impart a degree of unwholesome vulgarity to the wearer?
2. What did Harry Winston do with the rough piece of the 69.42 carat pear-shape diamond?
3. Did Burton give up purchasing the diamond? What did he do after the auction?
4. When did Taylor wear the diamond for the first time in public?
5. What is the present weight of the Taylor-Burton Diamond?

Subject Focus

1. Write a brief summary about the history of the Taylor-Burton Diamond.
2. Search for information on the Internet and find out more of the Taylor-Burton Diamond. Give a presentation in class.
3. Organize an auction show. You may be the auctioneer and bidders. Imitate the scenario in the text and act out the scene in class.

English for Jewelry

Language Focus

I. Subject-related Terms

Fill in the blanks with the words or expressions being defined.

1. _____ unit of weight (200 milligrams) for precious stones
2. _____ selling things in which each item is sold to the person who offers the most money for it
3. _____ someone who offers to pay a certain amount of money for something that is being sold
4. _____ a person who acts for, or manages the affairs of, other people in business, politics, etc.
5. _____ the money that has been obtained from a certain activity or an event

II. Working with Words and Expressions

In the box below are some of the words and expressions you have learned in this text. Complete the following sentences with them. Change the form of words if necessary.

outrageously	applicable	comment	display	acquire
enhance	erupt	previous	exclaim	simultaneous

1. Charges for local telephone calls are particularly _____.
2. It's the first time the painting has been _____ to the public.
3. He just _____ an antique painting.
4. Those clothes do nothing to _____ her appearance.
5. The theory does not seem easily _____ in this case.
6. His anger suddenly _____ into furious shouting.
7. You really can't _____ till you know the facts.
8. He could not help _____ at how much his son had grown.
9. The theatre will provide _____ translation in both English and Chinese.
10. Applicants for the job must have _____ experience.

III. Grammar Work

Observe the following sentences and pay special attention to the use of compound modifiers.

1. By the far the best known of Richard Burton's purchases was the *69.42-carat pear-shape*, later to be called the *Taylor-Burton* Diamond.
2. …the 50-year-old cleaver said nothing…that had separated from it and looked at it through his *horn-rimmed* glasses.
3. …speaking from a pay-phone of a *well-known* hotel in southern England, he spoke to…

Now correct the mistakes in the following sentences.

1. I was in a light-heart mood.
2. Miss Jones is a good-looked girl.
3. This has been a long-stand issue between two countries.
4. We have a shop selling cutting-price videos and CDs on Oxford Street.
5. Free-duty goods are sold at airports or on planes or ships at a cheaper price than usual.
6. Too many airlines treat our children as second-classes citizens.
7. People make a lastly-minute dash to the shops before Christmas day.
8. Try to behave in a more grow-up way.
9. A five-pages review of the novel is required in our literature course.
10. Undoubtedly, Wilson put forward a down-and-earth approach to solve the problem.

IV. Translation

1. Now this may be true of some women—usually wearing an outrageously large item of jewelry which imparts a degree of unwholesome vulgarity to themselves.
2. Markings were made, erased and redrawn to show where the stone could be cleaved.
3. Shortly afterwards on November 12th, Miss Taylor wore the Taylor-Burton in public for the first time when she attended Princess Grace's 40th birthday party in Monaco.
4. Miss Taylor announced that she was putting the diamond up for sale and was planning to use part of the proceeds to build a hospital in Botswana.
5. 这个规则对于他来说并不适用。
6. 到目前为止，哈里仍未就这些报道发表评论。
7. 它带来了至少 3 600 万美元的收益。
8. 现在正是你购买一套奢华的钻石项链的机会。

Part III Extended Learning

Dictation

Listen to the audio and fill in the blanks with words and phrases you hear.

The Philippine government will 1. _____ the huge collection of valuable pieces of jewelry confiscated from the family of late President Ferdinand Marcos in 1986, the Bureau of Customs (BOC) said Tuesday.

The BOC, which forged a Memorandum of 2. _____ (MOU) with the Presidential Commission on Good Government (PCGG), said they are committed to undertake successful physical inventory, 3. _____ and auction of the Marcos Jewelry Collection, which has an estimated value of 5 to 8 million US dollars, including the contested Malacanang Collection, according to a 4. _____ made in 1991.

The Marcos Jewelry Collection 5. _____ Roumeliotes Collection and Hawaii Collection, which the BOC holds in trust for the Republic of the Philippines.

The Roumeliotes Collection was 6. _____ from Demetrious Roumeliotes, an American citizen and an alleged close associate of the family of late President Ferdinand Marcos, in March 1986, while the Hawaii Collection was 7. _____ from the Marcos family when they were exiled to Hawaii.

Customs said an Inter-Agency Working Group (IWG) will be constituted to 8. _____ the efforts and responsibilities of the BOC, the PCGG and other concerned government offices to carry out the 9. _____ of auctioning the jewelry collection.

Proceeds from the auction would 10. _____ be remitted to the National Treasury, it added.

Read More

Passage 1

Modern Jewelry History

The European Renaissance was a benefit to all of the arts and innovation, and it opened the way for the Protestant Reformation that followed. We find the great artists of the time like Leonardo da Vinci starting their careers by working with well-known goldsmiths of the period. It is known that da Vinci, throughout his career, made jewelry designs for some of the prominent people who supported his artistic and inventive endeavors. All of the arts reached new levels of quality. The artisans of this time ushered in more modern approaches to the whole jewelry trade.

There was an increase in the brandishing of jewelry that spread all over Europe. Royal households attempted to outdo each other in every way, and that included which courts had access to the finest and most extravagant jeweled pieces. This type of competition was new, and had not been seen on such a level before.

Henry VIII of England had a great passion for jewelry, and this passion was carried on by his daughter, Elizabeth I. History says that Hans Holbein the Younger introduced Renaissance jewelry into England during the reign of Henry VIII. He was the primary designer of the monarch's personal jewelry. Henry had hundreds of rings produced, and he specifically liked jeweled hats.

Women of these periods often wore more than one necklace. This habit was characterized by the wearer donning a choker, and second longer necklace like a lengthy strand of pearls. Sometimes, a third prominent piece was worn as a clasp for a cape or other loose fitting garment.

The pendant seemed to be the most popular type of jewelry worn, and the technology used to produce them was superb. Some were classical cameos. Later gold and other gems were added to designs to give pendants a colorful, polychrome look. Polychrome was used to produce jewelry with a lot of different themes using openwork motifs and several linked items.

In Europe rings were manufactured in a large variety of styles. Certain rings could be opened and closed, and poison or other substances would be stored there. The storage of poison in rings worn by upper classes is legendary, and has served writers well when they fabricate mysteries.

On the heels of Renaissance artistic contributions, interest began to grow in Baroque art forms and it was especially prevalent in the 17th Century. This trend spread all over Europe. During this time artisans took the time to really view and study their craft and made improvements to it. There was a definite interest in floral art. Flowers became dominant themes for designers of fine jewelry. Other themes were used, but floral representations seemed to be the most popular.

The 17th Century marked an increase in the use of diamonds and other gemstones. Instead of wearing a bevy of jeweled items people started wearing stunning high quality pieces. The last monarch to wear jewelry to great excess was Louis XIV of France.

Brazil became a good resource for gem hunters who were looking for diamonds. Gem cutting became an important part of the whole jewelry trade, in large part, to the proliferation of diamonds.

Josiah Wedgewood was an English potter who ventured into the jewelry trade, and made significant contributions to the technology involved in jewelry production. He introduced oval, octagonal, and round plaques that was used for all types of jewelry.

The Industrial Revolution of the 19th century essentially created a jewelry market that was available to everyone. The middle class of a society could now purchases fine pieces of jewelry, and when imitation stones began flooding the market, even those from working classes could afford a piece of jewelry.

Commercial enterprises were formed to openly market the sale of jewelry. The firms of Faberge, Cartier, Tiffany and other great jewelry companies have their beginnings and roots in the Industrial Revolution. The Art Nouveau and abstract ideas such as Cubism with great artists like Salvador Dali and Pablo Picasso contributing to the trade.

Jewelry production does have a global history, and elaborate designs did appear in Asia, mainly in India and China. However, jewelry production never reached the heights that it did in Mid-Eastern, Egyptian, Classical and European traditions. Craftsmen in South America and parts of Mesoamerica were known for their work with gold and silver. The work of early South American and Mesoamerican goldsmiths can be found at the Museum of Gold in Bogota, Columbia and at historical museums in Mexico City.

Passage 2

The Queen's Jewels

The Queen's Jewels are a historic collection of jewels owned personally by the monarch of the Commonwealth realms; currently Queen Elizabeth II. The jewels are separated from the British Crown Jewels. The origin of a royal jewel collection distinct from the official crown jewels is vague, though it is thought that the jewels have their origin somewhere in the sixteenth century. Many of the pieces are from faraway lands and were brought back to the United Kingdom as a result of civil war, coups and revolutions, or acquired as gifts to the monarch.

The official Crown Jewels of the United Kingdom are worn only at coronations (St. Edward's

Crown being used to crown the monarch) and the State Opening of Parliament (the Imperial State Crown). On other formal occasions tiaras are worn. When the Queen goes abroad, she wears a tiara from her personal collection at formal events.

The Girls of Great Britain and Ireland Tiara

Queen Elizabeth II wearing the Girls of Great Britain and Ireland tiara

The Girls of Great Britain and Ireland Tiara was a gift from the girls of Great Britain and Ireland to the future Queen Mary in 1893. The diamond tiara was purchased from Garrard, the London jeweler, by a committee organised by Lady Eve Greville. In 1947, Mary gave the tiara to her granddaughter, the future Elizabeth II, as a wedding present.

The tiara was described by Leslie Field as "a diamond festoon-and-scroll design surmounted by nine large oriental pearls on diamond spikes and set on a bandeau base of alternate round and lozenge collets between two plain bands of diamonds".

Elizabeth II has usually worn the tiara without the base or pearls, but in recent years the base has been seen to have been reattached.

Over the years this tiara has become one of the most familiar of Queen Elizabeth II's tiaras through its appearance on British banknotes and coinage.

Princess Andrew of Greece's Meander Tiara

This tiara was a wedding gift to then Princess Elizabeth from her mother-in-law Princess Andrew of Greece and Denmark (born Princess Alice of Battenberg). The Meander Tiara is in the classical Greek "key pattern" featuring a large brilliant cut diamond in the centre surrounded by a diamond wreath. It also incorporates a central wreath of leaves and scrolls on either side. The Queen has never worn this item in public and it was given to Princess Anne around 1972. Princess Anne has frequently worn the tiara in public, notably during her engagement to Mark Phillips and for an official portrait marking her 50th birthday. In July 2011 the Princess Royal lent the Meander Tiara to her daughter Zara Philips for her wedding to Mike Tindall.

1936 Cartier Halo Tiara

This tiara was purchased by the Duke of York (later King George VI) for his wife (later Queen Elizabeth the Queen Mother) three weeks before they became King and Queen. It is a rolling cascade of scrolls that converge in a central ornament surmounted by a brilliant diamond. The tiara was then presented the future Queen Elizabeth II on the occasion of her 18th birthday.

The tiara was borrowed by Princess Margaret, before she was given the Persian Turquoise Tiara for her 21st birthday in 1951. Princess Margaret wore the Halo Scroll Tiara to the 1953 coronation of Queen Elizabeth II.

Queen Elizabeth II later lent the tiara to her daughter Princess Anne before giving her the Greek Meander Tiara in 1972.

The Halo Scroll tiara was lent to Catherine Middleton to wear at her wedding to Prince William on 29 April 2011.

 Group Project

Consult other references and share with your teammates the history of a famous jewelry.

Words and Expressions

Renaissance /rɪˈneɪsɑːns/	n.	文艺复兴
prominent /ˈprɒmɪnənt/	adj.	杰出的，卓越的
endeavor /ɪnˈdevə/	n.	努力
brandish /ˈbrændɪʃ/	vt.	挥舞
extravagant /ɪkˈstrævəgənt/	adj.	奢侈的
don /dɒn/	vt.	穿上，披上
choker /ˈtʃəʊkə/	n.	短项链
Baroque /bəˈrəʊk/	n.	巴洛克风格
coronation /ˌkɒrəˈneɪʃ(ə)n/	n.	加冕礼
proliferation /prəˌlɪfəˈreɪʃ(ə)n/	n.	扩散，增殖
octagonal /ɒkˈtæg(ə)n(ə)l/	adj.	八边形的
Cubism /ˈkjuːbɪz(ə)m/	n.	立体派
realm /relm/	n.	王国
tiara /tɪˈɑːrə/	n.	（女式）冕状头饰
surmount /səˈmaʊnt/	vt.	位于某物顶端
bandeau /ˈbændəʊ/	n.	（女用）束发带
lozenge /ˈlɒzɪndʒ/	n.	菱形
cascade /kæsˈkeɪd/	n.	波浪状物，花边
converge /kənˈvɜːdʒ/	vi.	相交，聚集
distinct from		与……截然不同

Unit 4

Jewelry Appraisal

Starting Out

☞ Match Words with Pictures

Match the words with the corresponding pictures.

| emerald   Chrysoberyl cat's eye   sapphire   tourmaline   rock crystal   Catier ruby |

1. _____   2. _____   3. _____

4. _____   5. _____   6. _____

☞ Check Your Knowledge

Fill the form with jewelry appraisal.

	RI (refractive index)	SG (specific gravity)	Dis (dispersion)	Natural (Y/N)	Hardness
diamond					
spinel					
tourmaline					
sapphire					
emerald					
chalcedony					

Part I Communicative Activities

Illustration of Products

Task 1 Conversations

I. Read the following conversation and underline some useful sentences for buying and selling jewelry.

(S: sales assistant; B: buyer)

S: Good afternoon. Can I help you?
B: I want a first-rate loose ruby.
S: What shape do you prefer? Round? Pear? Oval? Square? Heart?
B: I'm ready to make a pendant with 18K gold. What do you think?
S: Oh, pear-shape is the best, I think.
B: Could you choose one for me?
S: I recommend this one, madam. The ruby with vivid red and moderate grain size.
B: Thanks. Where is the ruby from?
S: Burma. With high quality and top-polish.
B: How much does it cost?
S: We sell it by carats. Wait a minute, please.
B: Sure.
S: 0.88 of a carat, $720.
B: Can you make it cheaper?
S: I'll give it to you for $700, OK?
B: We have to ask for another price reduction.
S: Well. $680, you can think about another cut.
B: I see, thanks a lot.

II. Look at the following conversation. Decide where the following sentences go and then act out the conversation in pairs.

a. it's tourmaline and the chain is 14K gold.
b. What is that?
c. I'd like emerald ring and pearl necklaces.
d. What's the price?
e. It's elegant and suitable for you.

(S: sales assistant; B: buyer)

English for Jewelry

S: Good morning. What can I do for you?
B: I'd like some jewelry.
S: All the jewelry is on sale today.
B: 1. _____
S: Sure. Here is a nice pearl necklace. 2. _____
B: May I have a look?
S: Yes, why not have a look at the nephrite bracelet and amethyst pendant by the way?
B: The necklace is very elegant. I'll take it. 3. _____
S: The gemstone? Oh, 4. _____
B: 5. _____
S: Its regular price is $880, and now you can have it with a twenty percent discount.
B: How about six hundred dollars?
S: I'm sorry we only sell at fixed prices.
B: Oh, I'll take the pearl necklace and that chain with fancy tourmaline.

Task 2   Role-play

Work in pairs and act out the following roles in the conversation about illustrating products.

Student A: a sales assistant in a jewelry store
Student B: a consumer wanting to buy a jewelry as a present for her mother

Useful Expressions and Sentences

Illustrating Products

- Our company is a professional manufacturer of platinum 900, karat gold jewelry, stones-setting with pearl, jade. We create 300 new designs per month with high quality, fast delivery and strict management. 本公司专业生产铂金PT900、K金饰品及各种镶嵌珍珠或翠玉的首饰, 每月推出300种新款式, 质量好, 交货快, 管理严谨。
- We serve the clients with perfect craftsmanship, fashionable style, competitive price and good service. 我们为客户提供做工精湛、款式新颖、价格合理、服务优质的产品。
- The company is well stocked with the largest collection of different cut diamonds at fair price. You are most welcome to contact us. 多种花形钻石货源充足, 价格合理, 欢迎广大客户光临。
- …has been well known in the world for fashionable design, unique cut and splendid craftsmanship……以新颖的设计、独特的造型以及精湛的工艺享誉海内外
- …is a professional processor in amber, coral, tusk and jewelry carving……是一家专业加工琥珀、珊瑚、象牙以及饰品雕刻的精品公司

Part II　Read and Explore

Gemstone Appraisal and Gemstone Value

　　Gemstone values tend to increase over time, particularly top-quality gemstones, which are scarce and always in high demand.

　　Although values tend to increase over time, there can be some fluctuation in the appraised values of any particular gemstone. To a large degree, this fluctuation is due to supply and demand, individual subjectivity, type of appraisal.

Supply and Demand

　　Both supply and demand can contribute to gemstone value fluctuations. An example of supply-side value increase would be a gemstone which has a limited geographic sourcing, such as tanzanite or tsavorite. If supply were to be interrupted (area and mine flooding, for example) and demand remained constant, it could be expected that the value of such a gemstone would increase significantly. The reverse could hold true as well, such as a new mine and source located with abundant supply for a gemstone that has previously been in scarce supply.

Individual Subjectivity

　　To an extent, the personal preferences of the appraiser can bear on appraised value of a gemstone. Cut, color, clarity, and carat weight essentially determine the desirability of a given gemstone variety. The weighting of each of these four variables can be influenced by preferences of the appraiser. For example, emeralds which are true-green in color may be graded and valued higher than those that are slightly bluish-green by a given individual appraiser, and the reverse may be true for a different individual appraiser.

Type of Appraisal

　　Replacement cost appraisals, commonly referred to as insurance appraisals, are designed to cover an individual in the event of loss of property, so these tend to result in higher valuations for gemstones and jewelry, because they are intended to provide coverage for replacement in a market that has fluctuating valuations.

　　Gemstone appraisal is an evolving science, and gemologists are in the midst of determining which standards would be best for gemstone evaluations. Many different types of gemstones and their colors, shapes, and sizes add the complexities of this process.

　　Currently there are no industry standards when it comes to valuing a gemstone (other than a diamond), but gemologists can still determine a good ballpark estimate for a gemstone based on the stone's 4Cs: cut, color, clarity and carat.

　　Although when it comes to determining a gem's value, the importance of each "C" can change dramatically depending on the type of gemstone being evaluated.

　　Below you will find helpful information about gemstone grades, certifications, appraisals and tips which you can use while shopping for gemstones and gemstone things.

The Importance of Cut in Colored Gemstones

Cut is perhaps the most consistently important C when it comes to determining the value of a gemstone. A poor cut, or a mediocre commercial cut, will fail to bring out the brilliance, fire, and color of a gemstone. A superior custom cut can be the difference between a pretty gemstone and a breathtaking gemstone.

For shoppers it can be difficult to immediately distinguish between a commercial cut and a superior custom cut, but taking the time to view different cuts of the same gemstone will give shoppers a clear idea of the differences between quality and mediocre cuts.

The Other 3Cs of Gemstone Value

Depending on the type of gemstone, color can make a huge difference in the gemstone's value. Take for instance the sapphire: a light yellow sapphire is priced on average around $500 per carat, but the rare and high in demand padparadscha sapphire with sunset and salmon colors is priced on average between $5,000 and $15,000 per carat.

The saturation of the gemstone's color will have an effect on gemstone appraisal too. Saturation is the intensity and depth of the gemstone's color, and gemstones with washed out or faded colors are much less desirable than those with a high degree of color saturation.

Clarity will also affect the value of a gemstone. Some gemstones, like emeralds, are formed naturally with inclusions, and as long as these inclusions are not visible to the eye they do not affect the stone's overall value. Yet with other types of gemstones, such as aquamarine, clarity is an important factor to determine the stone's value.

The effect that carat size has on a gemstone's value really depends on the type of gemstone being evaluated. The above mentioned aquamarine is an excellent example; this type of gemstone is often found in extremely large carat sizes, and so the carat size of an aquamarine has little to do with its value (a 40 carat aquamarine can cost no more than a 5 carat aquamarine). Other types of gemstones, like benitoite, which are rare and even rarely found in larger carat sizes, are valued largely by their carat sizes.

Gemstone Appraisal and Gemstone Certification

It is important to note that gemstone certification and gemstone appraisal are two different things.

A gemstone certification will grade the gem's clarity in addition to mapping the gem, and listing the gem's exact measurements, type of cut, and carat size.

A gemstone appraisal is offered after a certification has been issued for the gem. Many independent gemstone laboratories only offer certifications, but accredited independent gemologists will provide a certificate and an appraisal of the gem.

Check Your Understanding

I. **Fill in the blanks according to the text.**

1. The fluctuation in gemstone appraisal is due to _____, _____ and _____.
2. The stone's 4Cs are _____, _____, _____ and _____, which can be used by gemologists to determine the value of a gemstone.
3. It can be difficult for shoppers to immediately distinguish between a _____ and a _____.
4. If supply of a gemstone were to be interrupted, but demand remained constant, the value of the gemstone would _____ significantly.
5. A gemstone certification will grade the gem's _____, map the _____, and list the gem's _____, _____ and _____.

II. **Answer the following questions.**

1. What's the relationship between supply and demand in deciding gemstone's price?
2. How personal preferences of the appraiser determine the value of a gemstone?
3. Does replacement cost appraisal increase the value for gemstones or not? Why?
4. Which is the most important in determining a gem's value among the 4Cs?
5. What's the difference between gemstone appraisal and gemstone certification?

Subject Focus

1. Write a summary about the things that affect gemstone appraisal.
2. Go to a jewelry store and try to appraise certain gemstone, using the means you learnt from this passage.
3. Suppose your good friend wants to buy a gemstone present for his girlfriend/her boyfriend, please give him/her some advice on how to choose a real jewelry present, according to this passage.

Language Focus

I. **Subject-related Terms**

Fill in the blanks with the words or expressions being defined.

1. _____ rare, barely
2. _____ officially declared to be of an approved standard
3. _____ the process or state that occurs when a place or thing is filled completely with people or things, so that no more can be added
4. _____ developing gradually
5. _____ to change a lot in an irregular way
6. _____ to find out; to lie (in)

7. _____ the official or formal assessment of the strengths and weaknesses of someone or something
8. _____ of average quality
9. _____ suddenly and obviously
10. _____ judgment based on individual personal impressions and feelings and opinions rather than external facts

II. Working with Words and Expressions

In the box below are some of the words and expressions you have learned in this text. Complete the following sentences with them. Change the form of words if necessary.

| appraise | reverse | evaluation | wash out | overall |
| consistently | ballpark | intensity | faded | preference |

1. I can't give you anything more than just sort of a _____ figure.
2. He has demonstrated a strong _____ for being shod in running shoes.
3. With permanent tints, the result won't _____.
4. He is accused of _____ and callously ill-treating his wife.
5. The thoroughness of the _____ process we went through was impressive.
6. An employer should _____ the ability of his men.
7. For the country _____, house prices have remained flat.
8. The most visible sign of the _____ of the crisis is unemployment.
9. The company had to do something to _____ its sliding fortunes.
10. The books looked _____, dusty and unused.

III. Grammar Work

Observe the following sentences and pay special attention to the use of transitional relative clauses.

1. *Although* values tend to increase over time, there can be some fluctuation in the appraised values of any particular gemstone.
2. Currently there are no industry standards when it comes to valuing a gemstone (other than a diamond), *but* gemologists can still determine a good ballpark estimate for a gemstone based on the stone's 4Cs: cut, color, clarity and carat.
3. *Yet* with other types of gemstones, such as aquamarine, clarity is an important factor to determine the stone's value.

Now correct the mistakes in the following sentences.

1. Even we win tremendous successes in our work, we should not be conceited.
2. No matter what happens, but we will go on doing like that.
3. Although many independent gemstone laboratories only offer certifications, but accredited independent gemologists will provide a certificate and an appraisal of the gem.

4. While he was poor, he frequently invites others to dinner.
5. Even though the ruby brooch is expensive, but I won't hesitate to buy it for you.
6. I felt a bit tired. But I could hold on.
7. What you said was true nevertheless unkind.
8. Some of the studies show positive results, but whereas others do not.
9. In spite his illness, he clung tenaciously to his job.
10. She persevered in her idea despite of obvious objections raised by friends.

IV. Translation

1. A gemstone certification will grade the gem's clarity in addition to mapping the gem, and listing the gem's exact measurements, type of cut, and carat size.
2. Although values tend to increase over time, there can be some fluctuation in the appraised values of any particular gemstone.
3. Saturation is the intensity and depth of the gemstone's color, and gemstones with washed out or faded colors are much less desirable than those with a high degree of color saturation.
4. To an extent, the personal preferences of the appraiser can bear on appraised value of a gemstone.
5. 鉴定宝石时，可以先肉眼观察宝石的颜色、光泽、净度等特点。
6. 折射仪和天平是鉴定宝石的重要仪器。
7. 我们不仅要鉴别出宝石的种属，还要知道它是天然的还是合成的，是否经过人工优化或者处理。
8. 鉴定宝石时，根据所要鉴定的宝石的种类，每个"C"的重要性都会有很大变化。

Part III Extended Learning

Dictation

Listen to the audio and complete the following passage with the words or expressions you hear.

Cultural Relic Fair Opens in Beijing

The Beijing China Art International Fair has kicked off. Cultural relics from all over the world have been brought together under one roof, to be put on display, appraised and sold. The fair runs till next Tuesday, so interested buyers better 1. _____.

Protecting cultural relics by using the market is the aim of this fair. Gathering over one hundred art dealers, auction houses, and association 2. _____ from all over the world, the fair aims to create the largest platform to showcase and trade relics.

Kong Fanzhi, director of Beijing Administration of Cultural Heritage, said, "The value of cultural relics is realized through 3. _____. High prices draw higher public attention.

From this point of view, the antique market is also helping the protection of cultural relics."

Most of the cultural relics on display were brought by 34 time-honored antique companies from different provinces in China. In the coming four days, a series of special fairs will showcase 4. _____, Buddha statues, antique furniture and jewelry.

Participants will be offered free services from the organizer, including insurance, publicity, as well as a shuttle to and from the airport. Banks are also advertising their services at the fair.

Besides free entrance, visitors can attend seminars and have their jewelry appraised 5. _____. And it's all for free.

Yu Ping, deputy director of Beijing Administration of Cultural Heritage, said, "We are looking forward to seeing more and more 6. _____ and individuals to be part of the protection and collection of cultural relics. It is also a way of spreading knowledge and showcasing private collections."

The fair has also attracted overseas art dealers like Sotheby's and Christie's auction houses. Although their 7. _____ is just about over, they didn't want to miss any opportunity to get close to Chinese buyers.

The Association of Accredited Auctioneers is a collection of 19 auction houses from Britain. They are selling Chinese antiques back to China, which were brought to Europe during colonial times. Moreover, they see a 8. _____ market here.

Stephan Ludwig with Association of Accredited Auctioneers said, "But more importantly is looking ten, fifteen, twenty years forward, as China's new wealth, new middle class develop international taste in some of their collecting habits and decorating habits."

Last year, China raked in 30 percent of the world's 9. _____ of art and antiques, beating the US to become the number one market in the world. And 80 percent of that volume was in Beijing. That's more than 50 billion *yuan* a year. And every year, it has grown by a billion.

Reporter: "China boasts with 10. _____ cultural heritages. Given that advantage, Beijing wants to build itself as the world's fourth largest art transaction platform, after New York, London, and Hong Kong. With a booming market, the world starts to recognize the beauty and value of China's arts and culture."

Read More

Passage 1

Some Hints in Identifying Gemstone

Some hints could remind you of identifying gemstones when you observed the counterpart.

If any of the various optical phenomena are present—play of color, change of color and adularescence—the number of possibilities is reduced materially, weak asterism and chatoyancy are found in a number of species. Asterism is frequently seen in ruby, sapphire and orthoclase. Stones that show a cat's eye effect include the familiar chrysoberyl, quartz and tourmaline, but so do beryl, demantoid, nephrite, enstatite, diopside, feldspars, apatite, zircon, sillimanite and others.

A transparent, faceted stone that shows a red ring near the girdle when it is turned table-down on a white surface suggests a garnet-topped doublet. Flashes of red from a deep, vivid blue stone suggest synthetic spinel, or the tanzanite variety of zoisite.

Both zircon and synthetic rutile have exceedingly high birefringence, a condition easily recognized in transparent materials under magnification.

If there is strong doubling of opposite facet edges and the stone has natural inclusions, synthetic rutile is eliminated, and the unknown must be zircon. A doublet could be detected under magnification, or if there are bubbles and no doubling, glass. If the stone proved to be a diamond, however, only the spectroscope could distinguish between naturally and artificially colored material.

If the stone is doubly refractive with no visible inclusions, then specific gravity, strength of doubling, or strength of dispersion could distinguish between high-property zircon and synthetic rutile. Immersion in methylene iodide would show a great difference in refractive index by a great difference in relief.

Passage 2

Diamond—Not Just a Pretty Rock

Diamond are best known as the "girl's best friend" in rings, tiaras and the *Pink Panther* films. Yet the aesthetic uses for these sparklers—formed deep beneath the Earth's crust millions of years ago, at extremely high pressures and temperatures—are in the minority. Today, many diamonds are made synthetically for a vast range of uses, from surgeons' scalpels to super-fast microchips.

Labs produce 180 tonnes of diamond each year—almost nine times as much as comes out of the ground. And its strength, clarity and chemical resistance could make it the engineering material of the 21st century. Already, there are diamond heat sinks for tiny integrated circuits, diamond coatings on joint replacements and diamond windows on space probes. Soon, a diamond coating could protect your car gearbox, and super-strong diamond threads be used to reinforce ultra-light aircraft.

Diamond is far more than just a pretty rock. It has an impressive list of properties. As the hardest material known to science, it is resistant to attack by strong acids and alkalis and is a superb conductor of heat. This all means that there are many uses for diamond apart from the purely decorative. Because it dissipates heat so well—much better than silicon—engineers want to build microchips on layers of diamond. They could then squeeze yet more electronic components into smaller areas without the circuit overheating, to produce a new generation of super-fast computers. The key to diamond's extraordinary properties is its structure. Carbon can form four strong bonds with other molecules, which is why it forms the basis of so many organic compounds and is the building block of life. When four carbon atoms are linked together in a regular lattice, the result is a diamond crystal. Another form of carbon is the graphite in pencils.

Natural diamonds were formed up to three billion years of age in 200 km below the Earth's crust—in the mantle. They were then carried upwards in igneous rocks, such as kimberlite. As this molten rock approached the Earth's surface, it cooled to form the pipe structure in which natural diamonds are often found.

Companies such as De Beers and General Electric have been making synthetic diamonds since the early 1950s. Almost any substance rich in carbon can be converted into diamond. General Electric chemist Robert Wentorf once made diamonds from peanut butter.

A newer process, chemical vapor deposition (CVD), is used to produce ultra-hard diamond coatings. CVD used high temperatures but low pressures to coat a substance with carbon vapor, as a layer of small diamond crystals. These crystals will eventually join together and can be used to create huge gems. Diamonds such as this are usually sliced up to produce long scalpels or other tools.

CVD has scientists excited. "For the first time, we have all the superlative properties of diamond in a form that's useful for engineering applications," says May, who uses CVD to create diamond threads by coating tungsten wire. So it seems that diamonds are not just a girl's best friend, but an engineer's too.

Group Project

1. Interview some jewelry buyers and sellers, asking them how they identify the jewelries.
2. Compare the differences between the Chinese people and westerners in viewing and treating jewelries, and give a presentation on it.
3. Study the current development in the gemstone market and write a report about it.

Words and Expressions

appraisal /əˈpreɪz(ə)l/	n.	估计，估量，评价
scarce /skeəs/	adj.	缺乏的，罕见的
	adv.	勉强，刚，几乎不，简直不
fluctuation /ˌflʌktʃʊˈeɪʃ(ə)n/	n.	波动，涨落，起伏
appraise /əˈpreɪz/	vt.	估价，评估，估量，评价
subjectivity /ˌsʌbdʒekˈtɪvɪti/	n.	主观性，主观
supply-side /səˈplaɪˌsaɪd/	adj.	供应经济学政策的，通过减税而刺激生产和投资的
geographic /ˌdʒiːəˈɡræfɪk/	adj.	地理学的；地理的
tsavorite /ˈsɑːvəˌraɪt/	n.	铬钒钙铝榴石，沙弗莱石（透明、绿色的钙铝榴石，产于肯尼亚沙弗国家公园）
reverse /rɪˈvɜːs/	vt. & vi.	（使）反转；（使）颠倒

	adj.	相反的；颠倒的
	n.	相反，失败，挫折；倒转，反向
locate /ləʊˈkeɪt/	vt.	找出
	vt. & vi.	（在……）设置，坐落于
preference /ˈprefərəns/	n.	偏爱，爱好；优待，优先权；最喜爱的东西
appraiser /əˈpreɪzə/	n.	评价者，鉴定者，评估官
desirability /dɪˌzaɪərəˈbɪlɪti/	n.	愿望，希求
coverage /ˈkʌvərɪdʒ/	n.	新闻报道，报道量；提供的数量；覆盖范围（或方式）
fluctuate /ˈflʌktʃueɪt/	vi.	波动，涨落，起伏
evolving /ɪˈvɒlvɪŋ/	adj.	进化的，展开的
evaluation /ɪˌvæljuˈeɪʃ(ə)n/	n.	估价，评估
ballpark /ˈbɔːlpɑːk/	n.	棒球场
	adj.	大致正确的；大约的
dramatically /drəˈmætɪkli/	adv.	戏剧性地；引人注目地，显著地
tip /tɪp/	n.	小窍门，末梢，尖端
	vt.	倒掉，给小费
	vi.	翻转，倒翻
	adj.	倾斜的
consistently /kənˈsɪstəntli/	adv.	一贯地；坚持地，固守地
mediocre /ˌmiːdiˈəʊkə/	adj.	平庸的；普通的，平常的
padparadscha	n.	巴特帕拉德石，蓝宝石
salmon /ˈsæmən/	n.	鲑鱼，大马哈鱼；橙红色
saturation /ˌsætʃəˈreɪʃ(ə)n/	n.	浸湿，饱和；饱和度
intensity /ɪnˈtensɪti/	n.	强烈，剧烈；强度；烈度
faded /ˈfeɪdɪd/	adj.	已褪色的，已凋谢的
overall /ˌəʊvərˈɔːl/	adj.	总体的；全面考虑的
	adv.	大体上，全部的
	n.	工装裤
excellent /ˈeksələnt/	adj.	优秀的，卓越的；极好的
benitoite /bəˈniːtəʊaɪt/	n.	蓝锥矿；硅酸钡钛矿
accredited /əˈkredɪtɪd/	adj.	可接受的，公认的；官方认可的；获正式承认的
wash out		洗掉，破产，淘汰

Unit 5

Diamonds

Starting Out

☞ Match Words with Pictures

Match the words with the corresponding pictures.

| Star of Sierra Leone   Kohinur   Golden Jubilee Diamond   Great Star of Africa II |
| Great Star of Africa I   Great Mogul   Eureke   Woyie River Diamond |

1. _____ 2. _____ 3. _____ 4. _____

5. _____ 6. _____ 7. _____ 8. _____

☞ Check Your Knowledge

Fill the form with gemological characteristics of diamond.

Chemical composition	1.
Crystal system	2.
Mohs hardness	3.
Specific gravity	4.
Color	5.
Refractive index	6.
Dispersion	7.

Part I Communicative Activities

Inquiries, offers and counteroffer

Task 1 Conversations

I. **Read the following conversation and underline some useful sentences for buying and selling jewelry.**

(D: jewelry dealer; B: buyer)

D: Good afternoon, madam. May I help you?

B: I'd like to buy a diamond ring for myself.

D: Are you interested in any special brand?

B: No. Match is the most important, I think.

D: There are many styles of diamond ring and the prices vary from hundreds of dollars to thousands of dollars.

B: I don't know how you make the diamond's price.

D: Oh, You can think over the 4Cs, color, clarity, cut and carat weight.

B: Would you mind speaking more slowly?

D: Yes. You may look at these identification certificates. The grade of diamond is reported on.

B: I'm confused. You'd better find a diamond fitting me well. I believe you.

D: Platinum or K-gold?

B: I prefer nobel platinum.

D: Well, This is H-color, VS-clarity, excellent-round-brilliant cut, 1.02 ct platinum diamond ring. Would you try this one?

B: Oh, How nice! Do you think it'll look good on me?

D: Certainly, madam. It appears to be very exquisite and artistic.

B: Thanks a lot! I'll take it.

II. **Look at the following conversation. Decide where the following sentences go and then act out the conversation in pairs.**

> a. now you can have it with a twenty percent discount.
> b. I'm sorry we only sell at fixed prices.
> c. What's the price for this one?
> d. Is this one suitable for you?
> e. how much do you charge for it?

(D: jewelry dealer; B: buyer)

B: Do you have gold jewels?

D: Yes, we have 14K and 18K gold necklaces, chains and earrings.

B: May I have a look?

D: Sure. Here is a nice gold necklace. Its regular price is $56, and 1. _____
_____

B: It's very elegant. I'll take it. I want to buy some jewelry too.

D: What kind of jewelry do you like to have?

B: I would like to look at some bracelets.

D: Pure gold or carats?

B: Pure gold one, please.

D: Certainly, ma'am.

B: 2. _____

D: It's $650.

B: How about five hundred dollars?

D: 3. _____

B: I wish to buy a diamond ring, too.

D: 4. _____

B: No, it seems too old fashioned to me. Let me try it on. Oh, it's too small for me, haven't you got any larger ones?

D: How about this one?

B: This fits me well, 5. _____

D: Its regular price is $560.

B: Is that a real string of pearls?

D: You may take it on my word, if you find out it is an imitation you may return it to me.

Task 2 Role-play

Work in pairs and act out the following roles in the conversation about making offers and counteroffers.

Student A: a supplier of jewelry products located in Guangzhou

Student B: a buyer from Thailand

Useful Expressions and Sentences

1. Making an inquiry

- I'd like to buy a diamond ring for myself. 我想给自己买一枚钻戒。
- I don't know how you make the diamond's price. 我不知道钻石的价格是如何定的。
- Do you have gold jewels? 你有黄金首饰吗?
- What kind of jewelry do you like to have? 你想买哪类首饰呢?
- What's the price for this one? 这枚价格是多少?
- How much do you charge for it? 这个要价多少?

2. Making an offer

- I'm sorry we only sell at fixed prices. 对不起，我们是实价销售。
- There are many styles of diamond ring and the prices vary from hundreds of dollars to thousands of dollars. 钻石戒指有很多款，价格从几百美元到几千美元都有。
- Its regular price is $56, and now you can have it with a twenty percent discount. 它的原价是 56 美元，现在你可以 8 折买下。

3. Making counteroffer

- I find your prices are too high to be acceptable. 我发现你的价格太高了，我无法接受。
- I didn't expect that the discount you offer would be so low. 没有想到你给的折扣这么低。
- If you are able to make the price easier, I will take it. 如果你方出价更合理，我就会买下。

Part II Read and Explore

Gemological Characteristics of Diamond

Commercially, diamond is the most important of all gem species. It is estimated that diamonds account for approximately 90% of the value of gemstones purchased throughout the world. Diamond is always faceted to display its unique combination of adamantine luster and fire. Its supreme hardness ensures a lasting precision of cut which is unique among gemstones.

A common classification is industry diamond and jewelry diamond. The latter is usually graded by the "4Cs" criteria: carat, clarity, color and cut. In addition to the regular round, diamonds are often cut like pear, oval, heart, horse-eye, triangle, or emerald square. Elaborately designed, diamonds can be set to make all types of precious jewels such as necklace, earring, ring, etc. Gemological characteristics of diamond are listed below:

(1) Chemical composition: C (carbon).

(2) Crystal system: cubic.

(3) Habit: the most important is the octahedron. Diamond also occurs as tubes, dodecahedron, modified cubes. Crystals are frequently distorted, and crystal faces may be curved. Twinned octahedral crystals (macles are common).

(4) Surface features: triangular pits-trigons—may be seen on octahedral faces. Cleavage: perfect octahedral. Cleavage may be used in the fashioning of diamond, to split large crystals or trim off flawed material. Seen in and on cut and rough stones.

(5) Mohs Hardness: Diamond is the hardest known natural substance. The hardness of diamond varies according to the crystallographic orientation. But in any direction, diamond is still much harder than any other gemstone.

(6) Specific gravity: 3.52 g/cm^3.

(7) Color: colorless; yellowish, brownish or greenish. Fancy colors (those of a distinct

hue) include yellow and brown, rarely green, pink and blue, very rarely red and purple. Except for stones of fancy color, the desirability and value of diamond decreases as the depth of hue increases.

(8) Luster: adamantine.

(9) Refractive index: 2.42, single. Many diamonds display anomalous extinction.

(10) Dispersion: high, 0.044. Diamond displays a higher degree of dispersion than any other natural colorless gemstone.

(11) Luminescence: the fluorescence of diamond varies in color and intensity. It is stronger in long wave than short wave ultraviolet light. The color displayed may be bluish-white to violet, greenish or yellowish. Some stones are almost inert. The variability of diamond fluorescence is particularly useful in appraising jewelry set with many colorless stones. If all stones set in a piece show similar fluorescence, the stones are unlikely to be diamond.

Those diamonds which fluoresce blue under ultra violet light may show yellow phosphorescence. This is diagnostic for diamond.

(12) Occurrence: mainly from description of kimberlite pipes or alluvial deposits.

(13) Localities: the alluvial deposits of India were the only known source of diamond from classical times until the eighteenth century. The important Brazilian fields were discovered in about 1725.

The alluvial and kimberlite pipe deposits of South Africa were discovered in the latter half of the nineteenth century, and those in Siberia during the 1940s. More recently, Australia has become an important producer of diamonds which are found in lamproite, a rock similar to Kimberlite.

Important gem diamond producing countries include Angola, Australia, Botswana, Brazil, China, Namibia, Russia, Sierra Leone, South Africa, Tanzania.

(14) Imitations: many natural, synthetic and artificial products have been used to imitate diamonds. Of these, the most convincing, in terms of appearance, are: ① cubic zirconia (CZ); ② yttrium aluminate known as yttrium aluminium garnet or YAG; ③ colorless zircon; ④ some types of glass.

CZ is by far the best and most widely used simulant today. Other gemstones used as imitations include natural and synthetic white sapphire and synthetic white spinel.

Check Your Understanding

I. Fill in the blanks according to the text.

1. The jewelry diamond is usually graded by the "4Cs" criteria. The 4Cs are _____, _____, _____ and _____.
2. _____ is the universal measure of weight for a diamond.
3. Diamond clarity grade is classed into five categories: _____, _____, _____, _____ and _____.

4. _____ is the best and most widely used diamond imitation today in market.

II. Answer the following questions.

1. What is the hardness of diamond?
2. What is the meaning of "color" in diamond?
3. Which countries was the diamond produced?
4. What is the meaning of "imitations" in diamond?
5. What is "CZ"?

Subject Focus

1. Write a summary about Gemological Characteristics of Diamond and present it to the class.
2. Search for information on the "4Cs" criteria and give a presentation about how to distinguish between the diamond.
3. Suppose you are an online jewelry seller and you are putting some items of diamond on sale. Now make a draft of commodity description of a specific item, then post it online and present it to the class.

Language Focus

I. Subject-related Terms

Fill in the blanks with the words or expressions being defined.

1. _____ a heavy precious metallic element; grey-white and resistant to corroding; occurs in some nickel and copper ores and is also found native in some deposits
2. _____ something is extremely beautiful or pleasant, especially in a delicate way
3. _____ having or resembling repeated square indentations like those in a battlement
4. _____ a mark or flaw that spoils the appearance of something (especially on a person's body)
5. _____ when light travels through glass, it slows down and the amount it slows is measured by a number
6. _____ it is a crystalline form, twin-crystal or double crystal
7. _____ deviating from the general or common order or type
8. _____ light not due to incandescence; occurs at low temperatures
9. _____ it is a glow or soft light that is produced in the dark without using heat
10. _____ it is an amount of a substance that has been left somewhere as a result of a chemical or geological process

II. Working with Words and Expressions

In the box below are some of the words and expressions you have learned in this text. Complete the following sentences with them. Change the form of words if necessary.

| ensure | grade | distorted | decrease | extinction |
| appraise | trim off | in addition to | in terms of | by far |

1. In order to _____ success we must have a complete and thorough plan.
2. Analogously, a pattern, _____ or not, can be thought of as having a certain degree of energy.
3. Sales have now peaked, and we expect them to _____ soon.
4. Teachers always _____ their students.
5. We should _____ the connected roots of the mushrooms and rinse well.
6. Their area of the park—near the pizza boxes—is _____ the most dense.
7. The animals were ruthlessly hunted to the verge of _____.
8. How would you _____ the assignment?
9. _____ the "Top Ten Percent" program, the school considers race and other factors for admission.
10. Why do we have to define professionalism _____ basic supply and demand?

III. Grammar Work

Observe the following sentences and pay special attention to the use of passive voice.

1. It *is estimated* that diamonds account for approximately 90% of the value of gemstones purchased throughout the world.
2. The latter *is* usually *graded* by the "4Cs" criteria: carat, clarity, color and cut.
3. The important Brazilian fields *were discovered* in about 1725.
4. More recently Australia has become an important producer of diamonds which *are found* in Lamproite, a rock similar to Kimberlite.

Now correct the mistakes in the following sentences.

1. English was spoken by many people in the USA.
2. They was heard to sing (by us) in the next room.
3. The work will been finished soon.
4. A bike was been bought by him.
5. The work should being done carefully.
6. The flowers are be watered by them.
7. He paid the boy ten dollars for washing ten windows, most of them haven't been cleaned for at least ten months.
8. They were exciting at the story.
9. Dressing in white, he looks like a doctor.

10. The days were gone when China was both poor and backward.

IV. Translation

1. Commercially, diamond is the most important of all gem species. It is estimated that diamonds account for approximately 90% of the value of gemstones purchased throughout the world.
2. A common classification is industry diamond and jewelry diamond. The latter is usually graded by the "4Cs" criteria: carat, clarity, color and cut.
3. The variability of diamond fluorescence is particularly useful in appraising jewelry set with many colorless stones.
4. Many natural, synthetic and artificial products have been used to imitate diamonds. Of these, the most convincing, in terms of appearance, are: ① cubic zirconia (CZ); ② yttrium aluminate known as yttrium aluminium garnet or YAG; ③ colorless zircon; ④ some types of glass.
5. 除了彩色钻石外，钻石的价值随着色调的加深而降低。
6. 钻石的荧光变化范围较大，从无到强，荧光颜色多变。
7. 如果一件首饰上所有的宝石都呈现相似的荧光特征，则这些宝石就不大可能是钻石。
8. 到目前为止，合成立方氧化锆是最好的和应用最广泛的仿钻品。

Part Ⅲ   Extended Learning

 Dictation

Listen to the audio and complete the following passage with the words or expressions you hear.

Yuan Dips, Diamond Rises

At the end of last year, China became the world's second largest diamond consuming market. More Chinese people not only buy diamonds as a luxury good, but also hold them as an investment.

Cindy is looking for a 1.5 carat diamond. For her, a diamond is more than a piece of glamorous jewelry to wear.

"In addition to wearing them, diamonds are 1. _____ item now. Especially when the Renminbi is depreciating, jewelry can be an investment and also can be worn," Cindy said.

In the international market, the price of white diamonds has increased around 2. _____ in the past ten years. But in China, prices of some diamonds in retail stores have remained almost 3. _____ compared to a decade ago. Michael Huang from Diamond Index Group says that since diamond prices are US Dollar denominated, the gradual increase of the exchange rate of the Renminbi against the US dollar is part of the reason that 4. _____ in China have increased only slightly in recent years. But things have been different for the past several days.

"In the past several days, the Renminbi has 5. _____ the US dollar. The fall was around 2 percent a day. In this situation, the domestic prices of diamonds naturally increased. As far as I know, the prices of some diamonds have already increased 5 percent over the several past days," Huang said.

Some diamond retailers provide customers with investment plans based on diamonds, and the idea is that the retailers will repurchase them at a later date. They are looking 6. _____, however. For example, customers can buy diamonds from China Gold Group and then sell them back after a certain number of years. The price spread is the investment return. However, customers have to pay a total of 15 percent in transaction fees. According to the Shanghai Diamond Exchange, the average price increase of diamonds last year was around 10 percent. If that were the case, a two-year investment return on a diamond worth 100 thousand *yuan* would be around 2 percent, less than two-year 7. _____ at the bank. Xin Xiongfeng has been working in the jewelry industry for 12 years, and he says that since retailers are the end of 8. _____, it is unwise to invest in diamonds through them.

"Some diamond retailers have good branding, and promise good quality. But for diamonds, their qualities do not only rely on branding. If you do not buy your diamonds in retail stores, you avoid paying the cost of branding, and you will have a better return. That's why I do not recommend customers to buy diamonds for investment purposes in 9. _____," Xin said.

When investing in diamonds, customers should pay attention to 10. _____ —carat, color, clarity and cut. Industry insiders suggest that diamonds with more than 3 carats will hold their investment value better than smaller ones. China's trading volume of diamonds exceeded 43 million carats in the first four months of this year.

Read More

Passage 1

How to Clean Diamond Jewelry

Jewelry or ornaments such as necklaces, bracelets and earrings enhance one's beauty and charm. Though everybody likes dazzling jewelry, yet it is mostly the females who adore it. However, after wearing for a few days, the piece of bright and shining jewelry may lose its charm and glitter. Do you want to know why this happens? This is due to dust, dirt and the body lotions that we use. Jewelers often advise us to clean jewelry carefully on a daily basis. A good cleaning can turn a dull piece of jewelry into a charming and eye-catching one.

Before cleaning your jewelry, you should make sure of certain things such as the stones should be fitted properly. Make sure that all the clasps prongs, separation bicyclic ring, etc. that fasten the piece together are tight enough to clean. If you are not sure about how to clean your delicate jewelry piece, it is important to talk with your jeweler before using any homemade cleaning solution on it. Now, the question which comes to our mind is how to clean the jewelry. To answer this, let me tell you that there are various ways of cleaning jewelry. Jewelry can be cleaned at home too.

Tips for cleaning silver jewelry: a piece of silver jewelry should always be cleaned with soft cotton or flannel cloth.

* A silver cleaning cloth may help in quick cleaning your silver jewelry, as it has anti-tarnish ingredients.
* A smooth toothbrush can be used to clean the intricate scrollwork of the jewelry.
* A mixture of small amount of liquid detergent or soap in warm water can also used to clean dirt.
* Rinse the piece of jewelry in warm water, wipe with a cloth and allow it to dry.
* If you have used a toothbrush, scrub gently and rinse it.
* Use a silver dip (liquid cleaner) or baking soda paste to remove tarnish.

However, it is better to avoid using silver dip in cleaning gemstones, as the stones may get damaged due to chemical reactions.

Tips for cleaning gold jewelry: to clean a charming gold jewelry, such as necklaces, finger rings, bracelets studded with beautiful stones—soak the piece for 10 to 15 minutes in hot soapy water. Now scrub with a soft-bristle toothbrush, rinse it in lukewarm water and allow it to dry. Add a few drops of ammonia into the soapy water to remove the tarnish. In addition to this, dirt can be removed from your gold piece by dipping it into alcohol. In case of cleaning white gold jewelry, ultrasonic jewelry cleaning can be useful under the assistance of a jeweler.

Tips for cleaning diamond jewelry: every girl has a passionate desire to wear charming, precious diamond jewelry. Like other jewelry, it also may lose its shine after frequent wearing. To bring the sparkle back to your diamond jewelry, you should know how to clean the pieces. Here are some tips for cleaning your diamond jewelry:

* Soak your diamond earring, necklace, etc. in a solution of warm water and ammonia or mild detergent.
* Now scrub gently with a smooth toothbrush.
* A toothpick may help you to get hard-to-reach areas.
* Allow it to dry on a clean cloth or tissue paper.

While cleaning your diamond jewelry, don't use any homemade chlorine bleach as it may damage the charm and dazzle of the piece.

Tips for cleaning pearls: To clean jewelry made of delicate pearls, use soft and clean cloth and wet it in soapy water. Now rub the piece with this dampened cloth. Rinse it gently and allow it to dry. Do not soak your pearl jewelry in soapy water, as it may cause stretch. An enchanting pearl necklace, earrings, bracelet, etc. should be worn after cosmetics and perfumes are applied.

These jewelry cleaning tips may help you gain the shine of your adorable piece of jewelry. It is advisable not to use abrasive or other alcoholic mixtures while cleaning pearl jewelry, as alcohol may damage it. For better results or professional cleaning, you may approach a professional jewelry cleaner to polish your necklace or bracelet.

Passage 2

Understanding the Grading of the 4Cs of Diamonds

Grading Diamond Color

Since light source and background can have a significant impact on a diamond's appearance, diamond color is graded in a standardized viewing environment against color masters. A minimum of two color graders enter their independent evaluations into the system and depending on the agreement of these grades, and the weight and quality of the diamond, it may be sent to additional graders who enter their own color opinions. The grade is not determined until there is sufficient consensus.

Grading Diamond Clarity

Diamond clarity is graded under standard viewing conditions with 10x magnification. The preliminary grader carefully examines the diamond in order to identify clarity/finish characteristics and evidence of any clarity treatments such as fracture filling or laser drilling.

A minimum of two graders assigns their impression of the diamond's clarity, polish, and symmetry. Next they plot the clarity characteristics on the diagram most representative of the diamond's shape and faceting style.

Grading Diamond Cut

GIA provides a diamond cut quality grade for standard round brilliant diamonds that fall into the D-to-Z color range. To develop their Cut Grading System, GIA performed extensive computer modeling of round brilliant diamonds over a 15-year period and conducted more than 70,000 observations on actual stones to validate the research. This system can now predict the cut grade for more than 38.5 million proportion sets.

GIA's Excellent to Poor Cut Diamond Grading System assesses the diamond's overall face-up appearance to predict the intensity levels of brightness, fire, and scintillation (the diamond's sparkle and interplay with light). GIA also screens every diamond submitted to determine if it is synthetic.

Diamond Carat Weight Measurement

To determine diamond carat weight, the diamond is weighed using an extremely accurate electronic micro-balance that captures the weight to the precise fifth decimal place (the nearest ten-thousandth of a carat). An optical measuring device is used to determine the diamond's proportions, measurements, and facet angles.

Group Project

1. Choose one of the other important diamonds and write a passage about it, giving as many details as possible.

2. Compare the differences between the Chinese people and westerners in viewing and treating diamonds, and give a presentation on it.
3. Study the current development in the diamond market and write a report about it.

Words and Expressions

brand /brænd/	n.	商标，牌子，烙印
	vt.	打火印
confuse /kənˈfjuːz/	vt.	将……混淆，使糊涂
platinum /ˈplætɪnəm/	n.	白金，铂
exquisite /ɪkˈskwɪzɪt/	adj.	优美的，高雅的，精致的，剧烈的，异常的，细腻的，敏锐的
rutile /ˈruːtil/	n.	金红石
carbon /ˈkɑːb(ə)n/	n.	碳（元素符号 C）
graphite /ˈɡræfaɪt/	n.	石墨
indented /ɪnˈdentɪd/	adj.	锯齿状的，犬牙交错的
precise /prɪˈsaɪs/	adj.	精确的，准确的
brilliant /ˈbrɪli(ə)nt/	adj.	灿烂的，闪耀的，有才气的
crystallization /ˌkrɪstəlaɪˈzeɪʃ(ə)n/	n.	结晶化
birthmark /ˈbɜːθmɑːk/	n.	胎记，胎痣
proportion /prəˈpɔːʃ(ə)n/	n.	比例，均衡，面积，部分
	vt.	使成比例，使均衡
contribute /kənˈtrɪbjuːt/	v.	捐助，捐献，贡献，投稿
blemish /ˈblemɪʃ/	n.	污点，缺点，瑕疵
	vt.	弄脏，玷污，损害
extremely /ɪkˈstriːmli/	adv.	极端地，非常
colorimeter /ˌkʌləˈrɪmətə/	n.	色度计，色量计
elaborately /ɪˈlæbərətli/	adv.	苦心经营地，精巧地
octahedron /ˌɒktəˈhiːdrən/	n.	八面体
cube /kjuːb/	n.	立方体，立方
dodecahedron /ˌdəʊdekəˈhiːdr(ə)n/	n.	十二面体
distorted /dɪˈstɔːtɪd/	adj.	扭歪的，受到曲解的
macle /ˈmækl/	n.	双晶，斑点
triangular /traɪˈæŋɡjələ/	adj.	三角形的，三人间的
pits-trigons /pɪtsˈtraɪɡɒn/	n.	三角座
fashioning /ˈfæʃənɪŋ/	n.	精加工
anomalous /əˈnɒmələs/	adj.	不规则的，反常的
extinction /ɪkˈstɪŋkʃən/	n.	消光，消失，废止
luminescence /ˌluːmɪˈnesəns/	n.	发光
inert /ɪˈnɜːt/	adj.	无活动的，惰性的，迟钝的

phosphorescence /ˌfɑsfəˈresəns/	n.	磷光
kimberlite /ˈkɪmbərlaɪt/	n.	［地质］角砾云橄岩，金伯利岩（含金刚石）
alluvial /əˈluːviəl/	adj.	冲积的，淤积的
deposit /dɪˈpɒzɪt/	n.	堆积物，沉淀物，存款，押金，保证金，矿床
	vt.	存放
	vi.	沉淀
Siberia /saɪˈbɪəriə/	n.	西伯利亚
lamproite /ˈlæmprɔɪt/	n.	钾镁煌斑岩
Angola /æŋˈɡəʊlə/	n.	安哥拉
Sierra Leone /sɪˈerə liːəʊn/	n.	塞拉利昂
imitation /ˌɪmɪˈteɪʃ(ə)n/	n.	模仿，效法，冒充，赝品，仿造物
yttrium /ˈɪtriəm/	n.	［化］钇（稀有金属元素，符号 Y）
aluminium /ˌæləˈmɪniəm/	n.	［化］铝（符号 Al）
	adj.	铝的
YAG /jæɡ/		钇铝石榴石
chemical composition		化学成分
crystal system		晶系
Mohs hardness		莫氏硬度
flash effect		闪光效应
fracture filled diamond		裂隙充填钻石
metallic flux		金属熔剂
cloud inclusion		云状包裹体
laser drilling		激光钻孔
10x magnifying loupe		10 倍放大镜
a trained grader		训练有素的分级师
fancy colors		彩色
master stones		（钻石）比色石
adamantine luster		金刚光泽
twinned octahedral crystals		八面体双晶
fashioning of diamond		钻石的精加工
trim off		修剪
crystallographic orientation		结晶方向
ultraviolet light		紫外光
Kimberlite pipe		金伯利岩管
alluvial deposits		冲积物，冲积矿床

Unit 6

Ruby, Sapphire and Beryl (Emerald)

Starting Out

☞ **Match Words with Pictures**

Match the words with the corresponding pictures.

| garnet | beryl | star ruby | amethyst |
| padparadscha sapphire | aquamarine | | |

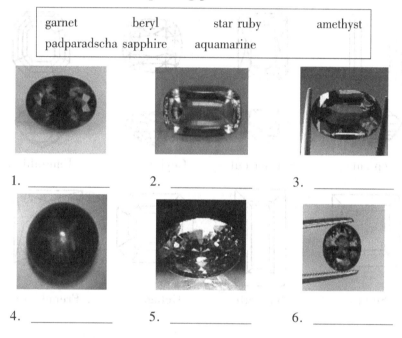

1. _____ 2. _____ 3. _____

4. _____ 5. _____ 6. _____

☞ **Check Your Knowledge**

Fill in the blanks with the types of cuts for gemstones.

There is no general rule which can be applied to the various cuts. However, three groups or types of cut can be named: faceted cut, plain cut and mixed cut.

The 1. _____ is practically applied only to transparent stones. The number of small even facets gives the gem higher luster and often a better play of color. Most facet cuts are built on two basic types, the brilliant cut and the trap or emerald cut. The 2. _____ can be leveled en cabochon (domed). This is suitable for agates and other opaque stones. In 3. _____, the upper part is level and the lower part is faceted, or vice versa.

Here are more types of cuts for gemstones.

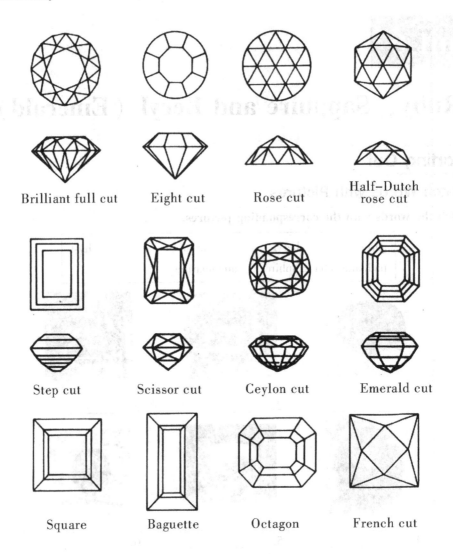

Part I Communicative Activities

Calling on Customer and Reception

Task 1 Conversations

I. Read the following conversation and underline some useful sentences for buying and selling jewelry.

A friend of Mr. Johnson is calling on a company to confirm a tour for the company.

A: By the courtesy of Mr. Johnson, we are given to understand the name and address of your company. If it is not too much trouble, we look forward to our tour for your firm, and moreover, we would like to talk to some of the technicians.

B: Thanks for your calling, OK, please let me know when you are free. We will arrange the tour for you in advance.

A: Thank you very much. I'll give you a call.

B: That's for sure, I believe that you will know our products better after the visit.

II. Look at the following conversation. Decide where the following sentences go and then act out the conversation in pairs.

> a. Very nice to meet you.
> b. I could use some coffee.
> c. How was the flight?
> d. you must be tired from the long flight.

(S: Vice President of Sales; M: Vice President of Marketing; G: General Manager)

S: Hello, Mr. Tanaka. Welcome to Silicon Valley Software. I'm Jack, VP of Sales.
   1. _____

G: I'm Takashi Tanaka, General Manager of Tokyo Technology. It's nice to meet you as well. Please call me Taka.

M: I'm Victoria, VP of Marketing. It's great to finally meet you face to face. After speaking so often on the phone, it's nice to "put a face with the voice". Thank you for coming all the way from Japan to visit our company. We've been looking forward to meeting you today.
   2. _____

G: It was long, but very good. I just arrived at San Francisco airport 3 hours ago.

S: Wow, 3. _____ Is the jet lag bothering you?

G: I'm fine. I was able to get some sleep on the airplane.

M: Would you like something to drink? We have coffee, tea, soda and water here.

G: Oh, thank you. 4. _____ The caffeine will help me stay awake during the meeting.

S: Of course. My assistant will bring it for you right away.

Task 2　Role-play

Work in pairs and act out the following roles in the conversation about calling on customer and reception.

Student A: the general manager of a company

Student B: the president of marketing

One is given a warm welcome by the other.

Useful Expressions and Sentences

1. I want to pay you a visit. Do you have spare time today? 我想过来看望你。你今天有空吗?
2. When will be a good time for me to pay you a visit? 我什么时候过来拜访你好呢?
3. Will it be convenient if I call you at 6:00 this evening? 我下午6点给你电话方便吗?
4. Could you please tell Mr. James I am here? 麻烦你通知詹姆斯先生我已经到了。
5. I'd like to see the person in charge of the sales. 我想见销售负责人。
6. I will be happy to wait. 我乐意等待。
7. I'm from ABC Company, and I am wondering if you would be interested in our products, perhaps you have heard our products before, would you like to know more about it? 我来自ABC公司, 你对我们公司产品感兴趣吗? 也许你曾经听说过我们的生产线, 你想了解更多吗?
8. I'm here to talk about the possibility of establishing business relations with your corporation. 我想就双方建立贸易关系的可能性和你探讨一下。

Part II　Read and Explore

Guide to Gemstones

　　Gemstones have played various roles in the myths and legends of human cultures throughout history. Some tell a story or are believed to have special powers, but all of them share a common beauty. Each gemstone is unique with a special color, birthplace and story. Gemstones come in every color of the rainbow and are gathered from all corners of the world, with each colored gemstone possessing a unique creation of beautiful color. Some gemstones have been treasured since before history began and others were only discovered recently. Join us as we explore the world of color gemstone jewelry.

Emerald: May Birthstone

Green is the color of Spring and has long symbolized love and rebirth. As the gem of Venus, it was also considered to aid in fertility.

Cleopatra, Egypt's tempestuous female monarch was as famous for wearing emeralds in her time as Liz Taylor is for wearing diamonds in our time. Ancient Egyptian mummies were often buried with an emerald carved with the symbol of verdure—flourishing greenness—on their necks to symbolize eternal youth.

The deeper and more vivid the color of green, the more valuable the gemstone. The most valuable and beautiful emeralds exhibit an intense bluish hue in addition to their basic bold green color. Emeralds, among the rarest of gems, are almost always found with birthmarks, known as inclusions. Some inclusions are expected and do not detract from the value of the stone as much as with other gemstones.

Ruby: July Birthstone

The Ruby represents love, passion, courage and emotion. For centuries this gem has been considered the king of all gems. It was believed that wearing a fine red Ruby bestowed good fortune on its owner. Rubies have been the prized possession of emperors and kings throughout the ages. To this day the Ruby is the most valued gemstone.

The color of a Ruby is the most important feature of the gemstone. Rubies are available in a range of red hues from purplish and bluish red to orange-red. The brightest and most valuable color of Ruby is often "a Burmese Ruby" —an indication that it is a rich, passionate, hot, full red color with a slight blue hue. This color is often referred to as "pigeon blood" red, a Ruby color only associated with the Mogok Valley mines in Myanmar. The color Pigeon Blood Ruby red, is not associated with the color of a pigeon's blood but rather the color of a white pigeon's eye.

Sapphire: September Birthstone

When hearing the word "sapphire" many people immediately envision a stunning violet-blue gemstone because the word "sapphire" is Greek for blue. For centuries, the sapphire has been referred to as the ultimate blue gemstone. Since Ancient times the Blue Sapphire represented a promise of honesty, loyalty, purity and trust. To keep with this tradition sapphires are one of the most popular engagement gemstones today.

Sapphire is found in many parts of the world, but the most prized sapphires are from Myanmar (Burma), Kashmir and Sri Lanka. Sapphires with highly saturated violet-blue color and "velvety" or "sleepy" transparency are more rare. The purer the blue of the sapphire, the greater the price. However, many people find that the darker hues of sapphire can be just as appealing.

Sapphires are not only blue, they come in almost every color of the rainbow: pink, yellow, orange, peach, and violet colors. The most sought-after color fancy sapphire is the rare and beautiful Padparadscha: a pink-orange corundum with a distinctive salmon color reminiscent of a tropical suns.

English for Jewelry

Check Your Understanding

I. Fill in the table with the given information in the text.

	Emerald	Ruby	Sapphire
Birthstone			
Color			
Symbol			
Status			
Important features			

II. Answer the following questions.

1. According to the text, people from which country were most famous for the love of emeralds? Why?
2. Why do not some inclusions in emeralds detract from the value?
3. Why is ruby considered the king of all gems?
4. What is "pigeon blood" associated with?
5. How does the color of gemstones influence their price?

Subject Focus

1. Choose one of the gemstones in the text and design an advertisement with slogans and pictures.
2. Search for information on the Internet and find out more about other gemstones. Give a presentation in class.
3. A jewelry fashion show is coming. Design jewelry with gemstones and write out a short script describing your jewels and present it when models display jewels.

Language Focus

I. Subject-related Terms

Fill in the blanks with the words or expressions being defined.

1. _____ is a characteristic enclosed within a gemstone, or reaching its surface from the interior
2. _____ is one of the main properties (called color appearance parameters) of a color, the degree to which a stimulus can be described as similar to or different from stimuli that are described as red, green, blue and yellow

3. _____ is the colorfulness of a color relative to its own brightness

II. Working with Words and Expressions

In the box below are some of the words and expressions you have learned in this text. Complete the following sentences with them. Change the form of words if necessary.

| bestow | intense | indication | distinctive | ultimate |
| eternal | symbolize | stunning | envision | reminiscent |

1. A red sky at night _____ fine weather the following day.
2. My experience as player, coach and manager has prepared me for this _____ challenge.
3. The Queen has _____ a precious ruby on him.
4. Why does pigeon _____ peace?
5. You look really _____ in that dress!
6. In the future, we _____ a great progress in the field of artificial intelligence.
7. We'll be _____ grateful to our family.
8. This painting is strongly _____ of da Vinci's *Annunciation*.
9. He gazed at me with those _____ blue eyes.
10. The shell has a highly _____ pattern.

III. Grammar Work

Observe the following sentences and pay special attention to the suffix "-ish".

1. The most valuable and beautiful emeralds exhibit an intense *bluish* hue in addition to their basic bold green color.
2. Rubies are available in a range of red hues from *purplish* and *bluish* red to orange-red.

Now rewrite the following sentences using the words with "-ish".

1. He knew what he had done was immature, just the way like a child.
2. She had slightly red and brown hair.
3. Dan is a smart, dark-haired and fairly young man.
4. She gave a little giggle, shy like a small girl.
5. It is commonly believed that people tend to be self-centered.

IV. Translation

1. Gemstones come in every color of the rainbow and are gathered from all corners of the world, with each colored gemstone possessing a unique creation of beautiful color.
2. Cleopatra, Egypt's tempestuous female monarch was as famous for wearing emeralds in her time as Liz Taylor is for wearing diamonds in our time.
3. The deeper and more vivid the color of green, the more valuable the gemstone.
4. Since Ancient times the Blue Sapphire represented a promise of honesty, loyalty, purity and trust.

5. 此行情必然对细分市场形成起到推进作用。

6. 绿色象征着爱和重生。

7. 祖母绿一般都会存在内含物，而这并不会像其他宝石一样损害宝石的价值。

8. 红宝石色调涵盖了紫红色和蓝红色到橙红色的红色色调。

Part Ⅲ Extended Learning

Dictation

Listen to the audio and complete the following passage with the words or expressions you hear.

HONG KONG, Feb. 20 (Xinhua) —Two key jewelry shows will open in Hong Kong next week, forming the largest jewelry 1. _____ in the world.

The 4th Hong Kong International Diamond, Gem and Pearl Show will open next Tuesday and 2. _____ March 4 at the Asia World-Expo.

The 34th Hong Kong International Jewellery Show will open next Thursday and run through March 6 at the Hong Kong Convention and Exhibition Center.

The twin shows, 3. _____ by the Hong Kong Trade Development Council (HKTDC), has attracted a record of 4,480 4. _____ from 53 countries and regions.

Despite global economic uncertainties, the strong participation of exhibitors and buying missions at the shows 5. _____ that the jewelry industry remains optimistic about business prospects and sees Hong Kong as an important sourcing platform for jewelry and 6. _____, said Benjamin Chau, HKTDC deputy executive director.

The Hong Kong International Diamond, Gem and Pearl Show features three signature zones, including one for diamond suppliers, another for precious 7. _____, and the third for pearls.

The jewelry show will gather 8. _____ local and global jewelry brands and new designer brands.

Jewelry 9. _____ demonstrations and themed seminars will be held to help promote the latest market information and crafts and product 10. _____.

Read More

Passage 1

Twenty Fun Emerald Facts

1. Emerald is one of the four recognized precious gemstones. The others are ruby, sapphire and diamond.

2. Top quality emeralds can be worth more than diamonds because emeralds without imperfec-

tions are very rare.
3. A 1 carat emerald appears larger than a 1 carat diamond because emeralds have a lower density.
4. Most emeralds have some type of inclusion or imperfection. Instead of use the term imperfection, dealers like to reference emerald inclusions as an internal "jardin" (garden in French).
5. Most emeralds are treated usually by filling the emerald with an oil to fill in the cracks and help prevent unintentional chipping or cracking.
6. Due to the inclusions within an emerald, it is not wise to clean these gems in an ultrasonic cleaner. Instead clean gently by hand using warm water.
7. Emerald is made of beryl just like aquamarine but gets its green coloring from very small amounts of chromium and/or vanadium.
8. Color, clarity, cut, and carat weight are four factors used to determine the value of an emerald. The most important of these four is color. The best color is vivid green or blueish green with even saturation and no color zoning. It is also important that the emerald is very transparent and isn't too dark or too light.
9. The oldest emeralds are about 2.97 billion years old.
10. The first known emeralds were mined in Egypt around 1500 BC.
11. One of Cleopatra's favorite stones was emerald.
12. Emeralds were discovered in South America in the 16th century by the Spanish. They were used by the Incas well before this discovery.
13. The Spanish then traded these emeralds across Europe and Asia for precious metals, opening up the emerald trade to the rest of the world.
14. According to ancient folklore, putting an emerald under your tongue would help one see into the future.
15. Today, Colombia yields the largest amount of emeralds, contributing to more than 50% of all emerald production worldwide.
16. The Duke of Devonshire Emerald is one of the largest uncut emeralds weighing 1,383.93 carats.
17. Synthetic sapphire and ruby were created in 1907, but synthetic emeralds were not created until 1935 when American chemist Carroll Chatham successfully grew his first one carat Chatham emerald which is now on display at the Smithsonian Institute.
18. Emeralds were first discovered in North America in the Yukon Territory in 1997, though large emerald deposits in the United States and further north are very rare.
19. A gemologist judges a diamond's clarity grade by using a 10x loupe. The clarity of an emerald is often assessed with the naked eye.
20. An emerald pendant necklace owned by Elizabeth Taylor sold for $6.5 million in 2011, breaking down to about $280,000 paid per carat.

Passage 2

Tips on Buying Gemstones

Gemstones have been sought after and treasured throughout history. They have been found in ruins dating several thousand years. They are valued as gifts symbolizing love.

Generally, the price of any gemstone is determined by: size, cut, quality color/clarity/treatments and type. Here are some questions to ask about quality:

- Has it been treated? (See treatments listed below)
- Is the stone natural or synthetic?
- Are there any noticeable scratches, chips or inclusions?
- Is the color even throughout the stone?
- How good is the color? (Is it vivid?)
- If you are buying the stones for earrings or cufflinks, are the stones well matched?

There are many ways that dealers treat gemstones. The savvy buyer asks lots of questions and hopefully tests the results. Here are some treatments to look for:

- Irradiation: It is common to irradiate aquamarine, London blue topaz, emerald, and diamond as well as other stones. This treatment brings out color and removes imperfections. Many dealers know if the stones they are selling have been irradiated. Honest ones will tell you if they are aware of the treatment.
- Heat Treatment: amethyst, aquamarine, ruby, tanzanite and topaz are often heated at high temperatures to enhance color.
- Dye: This is the most common treatment used. On clear stones, dye may be visible in cracks that are darker than the rest of the stone. Sometimes dye appears as a residue that rubs off or white patches. Lapis and rose quartz are commonly dyed. Amethyst and citrine are often dyed. Black onyx is permanently dyed in normal processing.
- Coatings: Jasper is often dipped in petroleum products to bring out color and to seal it. Emerald is oiled; turquoise is waxed.

Tips on buying beads:

- Good sized holes (so can use a stronger thread).
- Evenly shaped beads (as appropriate).
- If the beads are being sold in a 16 inches strand—not 14 inches or 15 inches if possible.
- Look for the best quality stone (if buying real stones).
- Make sure beads are not cracked or chipped by the holes as this tears the thread.
- Good color (so can create harmonious necklaces and matching earrings).

Group Project

1. Choose one of your favorite gemstones and introduce it to your group members.
2. Compare the differences between the Chinese people and westerners in viewing and treating emeralds, and give a presentation on it.

3. Study the current development in the emerald market and write a report about it.

Words and Expressions

imperfection /ˌɪmpəˈfekʃ(ə)n/	n.	瑕疵，缺点
density /ˈdensɪti/	n.	密度
inclusion /ɪnˈkluːʒ(ə)n/	n.	包含物
crack /kræk/	n.	裂缝
chip /tʃɪp/	n.	小缺口
beryl /ˈberɪl/	n.	绿宝石
aquamarine /ˌækwəməˈriːn/	n.	海蓝宝石
chromium /ˈkrəʊmɪəm/	n.	铬
vanadium /vəˈneɪdɪəm/	n.	钒
transparent /trænˈsper(ə)nt/	adj.	透明的
folklore /ˈfəʊklɔː/	n.	民间传说
synthetic /sɪnˈθetɪk/	adj.	合成的
scratch /skrætʃ/	n.	划痕
irradiation /ɪˌreɪdɪˈeɪʃ(ə)n/	n.	辐照，照射
topaz /ˈtəʊpæz/	n.	黄玉
amethyst /ˈæməθɪst/	n.	紫水晶
dye /daɪ/	vt.	染色
residue /ˈrezɪdjuː/	n.	剩余物，残余
lapis /ˈlæpɪs/	n.	天青石
coating /ˈkəʊtɪŋ/	n.	涂层
turquoise /ˈtɜːkwɔɪz/	n.	绿松石
wax /wæks/	vt.	上蜡
bead /biːd/	n.	珠子
verdure /ˈvɜːdʒə/	n.	碧绿，翠绿
bluish /ˈbluːɪʃ/	adj.	浅蓝的，有点蓝的
hue /hjuː/	n.	色调，色度
purplish /ˈpɜːplɪʃ/	adj.	略带紫色的
indication /ˌɪndɪˈkeɪʃ(ə)n/	n.	指示，迹象
envision /ɪnˈvɪʒ(ə)n/	vt.	想象，预想
stunning /ˈstʌnɪŋ/	adj.	极好的
reminiscent /remɪˈnɪs(ə)nt/	adj.	使人回想或联想起的
Myanmar /ˌmɪænˈmɑː/	n.	缅甸
associate with		将（人或事物）联系起来
sought after		受追捧的
detract from		使……的价值或重要性受损、减少
in addition to		除……以外（还有）

Unit 7

Polycrystalline Gemstones

Starting Out

☞ **Match Words with Pictures**

Match the words with the corresponding pictures.

| jadeite   nephrite   Dushan jade   turquoise   serpentine jade   quartzite |

1. _____ 2. _____ 3. _____

4. _____ 5. _____ 6. _____

☞ **Check Your Knowledge**

Fill in the blanks with details about the different aspects of the following types of gemstones.

Type / Item	Jadeite jade	Nephrite	Dushan jade	Turquoise	Serpentine jade	Quartzite
Chemical composition	$NaAl(SiO_3)_2$, Sodium, aluminum silicate					
Hardness	About 7					

80

续上表

Type / Item	Jadeite jade	Nephrite	Dushan jade	Turquoise	Serpentine jade	Quartzite
Luster	Greasy to vitreous					
Transparency	Transparent (very rare), translucent to opaque					
Specific Gravity (SG)	3.30 – 3.36					
Refractive index (RI)	1.65 – 1.67					
Colors	White, shades of mauve, violet, red, orange, yellow, brown, pale to deep emerald green, deep green to black					
Localities	Burma, Kazakhstan, Japan, California USA					

Part I Communicative Activities

Inquiry, Offer and Counteroffer

Task 1 Conversations

I. **Read the following conversation and underline some useful sentences for buying and selling jewelry.**

(D: jewelry dealer; B: buyer)

D: Welcome to our company, Mr. Smith. Take a seat, please.
B: Thank you.
D: I've been told that you showed an interest in some of our products displayed at the Fair.
B: Yes, your products are very impressive.
D: We're glad to hear that. What items are you particularly interested in?
B: I've seen the exhibits and studied your catalogues. We want to buy some jadeite jade ornaments. Here's a list of our requirements. Could you give me some idea of the prices?
D: Here is our latest price sheet. You will see that our prices are the most competitive.
B: These are all FOB prices. Could you please quote us the prices CIF Phoenix?
D: Would you tell us the quantity you require so that we can work out the offer?
B: Yes, we want 1,000 pieces of the new products.
D: We'll have them worked out by this evening and make the offer tomorrow morning. Will you be free to come by then?
B: OK, I'll be here tomorrow morning at 10.

II. **Look at the following conversation. Decide where the following sentences go and then act out the conversation in pairs.**

> a. Your price compares unfavorably with your competitors.
> b. I suppose we could consider giving you a 4% discount.
> c. What price would you quote us on the items on our inquiry sheet?
> d. I would like to make you a special offer.
> e. What's your idea of a competitive price?
> f. Here's our offer, $100 per piece, FOB Guangzhou.

(S: supplier; B: buyer)

S: Good morning, Mr. White. It's nice to have you here.
B: Good morning, Mr. Chen. I've checked the catalogues and seen some of the samples. We are particularly interested in the jade ornaments. 1. _____

S: Well, 2. _____ You will notice the quotation is much lower than the current market price.

B: I'm afraid I can't agree with you there. We have quotations from other sources too. 3. _____

S: Well, then, 4. _____

B: As we do business on the basis of mutual benefit, I suggest somewhere around $90 per piece FOB Guangzhou.

S: I'm sorry the difference between our price and your counteroffer is too wide. Our jade ornaments are of high quality and unique design. If you increase your initial order to 3,000 pieces, 5. _____

B: I'm afraid it's too large a number for a trial order. How's this, 2,000 pieces at $95 per piece?

S: All right, considering the expanding market, 6. _____

B: I'm glad we have brought this transaction to a successful conclusion.

Task 2　Role-play

Work in pairs and act out the following roles in the conversation about making offers and counteroffers.

Student A: a supplier of jewelry products located in Guangzhou

Student B: a buyer from Thailand

Useful Expressions and Sentences

1. **Making an inquiry**
- Can you give me an indication of price? 你能给我一个估价吗？
- I would like to make an inquiry about this type of... 我想询问一下这种……的价格。
- What price could you quote us on...? 你们给我们报的……价格是多少？
- Would you please quote us the lowest price for...? 能否报给我们……的最低价？
- We want to know the price CIF...（place）for your...（product）. 我们想知道……（产品）的……（地点）到岸价。

2. **Making an offer**
- You will see that our prices are the most competitive. 你会发现我们的价格是最具竞争力的。
- We can quote you the price of...per piece. 我方的报价是每件……元。
- This is the best price we can give you. 这是我报给你的最优价格。
- The price we quoted is on a FOB Shanghai basis instead of a CIF Bombay basis. 我方的报价为上海离岸价，而不是孟买到岸价。
- If you increase your initial order to 1000 pieces, I suppose we could consider giving you a 4% discount. 如果你将首单增加到1000件的话，我想我们可以给你们4%的折扣。
- We cannot make any further discounts. 我们不能再给任何的折扣了。

3. Making counteroffer

- I'm afraid the offer is unacceptable. 恐怕你方的报价我们不能接受。
- We can't accept your offer unless the price is reduced by 5%. 除非你们减价5%，否则我们无法接受你们的报价。
- I'm afraid I don't find your price competitive at all. 我看你们的报价毫无竞争力。
- If you hang on the original offer, business is impossible. 如果你方坚持按原来的报价，生意根本没办法谈下去。

Part II   Read and Explore

Chinese Jadeite Bangle: White Translucent Jade

Item: White jadeite bangle

Of semi cylindrical form.

White tone of good consistency throughout the bangle with semi translucency.

Tested to be genuine jadeite jade, test results below.

Age, condition, provenance and dimensions are stated below the test results.

We reserve the right to end this listing.

More information can be found below in sale information.

Our specialist has tested the jadeite bangle in the following aspects:

1. With the thermal conductivity test which shows it as jadeite.

2. With a polariscope and the jadeite bangle is polycrystalline of which is that of jadeite.

3. With a Chelsea filter and there is no red reaction, which suggests there is no color treatment.

4. With forensic level UV lighting and when subjected to the ultraviolet radiation the bangle does not release any ultraviolet light, which is a sign of natural grade A jadeite.

5. With magnification and the bangle has the structure and characteristics of jadeite.

6. With a hardness test and the bangle measures 6.5 which is the hardness of jadeite.

7. With a basic spectrometer and the absorption lines are in the areas representative of natural grade A jadeite.

8. The craftsmanship and polish of the bangle is of high quality, which separates the bangle from any low quality bangle.

To conclusively test for dyed or treated jadeite for the most sophisticated treatments, highly expensive equipment is needed such as a Raman spectrometer and a Fourier transform infrared spectrometer.

To be safe, our specialist recommended to offer the bangles for sale as genuine jadeite as they have been tested to be, although to be safe not to guarantee them as grade A as further tests with more advanced equipment would be needed to conclusively give a guarantee.

To conclude the bangle is jadeite and is possibly grade A as the tests undertaken show a significant probability of grade A.

After we upgrade our testing facility we will start to provide a grade A guarantee on jadeite although until then we are just being safe and cautious.

We notice many sellers labeling items as grade A without doing any significant testing and beware of sellers that have copied our information.

All our test results can be backed up and shown to whom it may concern with pictures if necessary.

Age: most probably mid to late 20th Century.

Provenance: property of a distinguished private collector of Chinese heritage.

This piece forms part of the Chinese jewelry selection from the same Australian estate collection we had received items from previously and had listed some nephrite jade pieces from last Sunday week.

The overall collection consists of mostly Chinese antiques, amassed through many years of traveling and collecting.

Condition: in very good condition, no damage or repairs.

This piece contains natural inclusions and variations and minor natural fractures.

Dimensions:

Outer diameter: 69 mm (2.7″)

Inner diameter: 54 mm (2.1″)

Width: 15 mm (0.6″)

Measurements may vary slightly.

Sale information

Adelaide Asian Auction House is pleased to offer this Jadeite bangle from a remarkable selection of works of art comprised of various private collections in Australia, consisting of jades, bronzes and other items of mainly Chinese origin.

The collections we obtained recently were amassed with great care over many generations and generally the collections were purchased by the collectors from all over the world during their travels.

This particular item is to be offered online. On behalf of the owner or estate, we reserve the right to withdraw the item from sale if unexpected events occur. If such an event occurs we will offer a 5% – 10% discount on any future single purchase upon mention of the particular occurrence.

Jade info

There are two types of Jade, nephrite and jadeite. Nephrite is much more common.

Nephrite first appeared in China about 5000 BC but jadeite was not documented to appear in China until the mid 18th century when large deposits were imported into China. Jadeite is slightly

English for Jewelry

harder than nephrite and is often used for jewellery.

Jadeite is more popular and generally much more valuable by weight than nephrite.

Prices of quality raw jade and especially jadeite have skyrocketed since 1980.

Market price of raw white river jade (not Jadeite) historically has been:

1980: 100 *yuan* per kg

2000: 30,000 *yuan* per kg

2010: 300,000 *yuan* per kg

And rarer, more sought after raw pieces were typically selling for 2,000,000 *yuan* per kg in 2010.

Check Your Understanding

I. **Fill in the following table about the identification of the Chinese jadeite bangle according to the text.**

Item Tested	Testing Means	Result of Test
Thermal conductivity		
Crystallographic structure		
Color		
Luminance		
Internal structure		
Hardness		
Absorption spectrum		
Craftsmanship		
Conclusion		

II. **Answer the following questions.**

1. Why did the specialist choose not to guarantee the bangle as grade A jadeite?
2. What do you think of the seller in terms of identifying the bangle?
3. How do you value the bangle according to the information given in the text?
4. What are the differences between nephrite and jadeite?
5. What is the trend in the jade market?

Subject Focus

1. Write a summary about the Chinese jadeite bangle and give a presentation to the class.
2. Search for information on the grades of jadeite jade and give a presentation about how to distinguish the four different grades of jadeite jade.
3. Suppose you are an online jewelry seller and you are putting some items of nephrite jade on sale. Now make a draft of commodity description of a specific item, then post it online and present it to the class.

Language Focus

I. Subject-related Terms

Fill in the blanks with the words or expressions being defined.

1. _____ the quality of allowing light to pass diffusely
2. _____ where something originated or was nurtured in its early existence
3. _____ the magnitude of something in a particular direction (especially length or width or height)
4. _____ composed of aggregates of crystals
5. _____ the process of cutting, polishing and enhancing gemstones to improve their appearance, durability or availability
6. _____ the act of expanding something in apparent size
7. _____ the quality that a gemstone has when it is beautiful and has been very carefully made
8. _____ the property of being smooth and shiny
9. _____ any irregularity observable in a gem by the unaided eye or some tool
10. _____ a slight crack or break in gemstones

II. Working with Words and Expressions

In the box below are some of the words and expressions you have learned in this text. Complete the following sentences with them. Change the form of words if necessary.

| reserve | release | measure | separate | genuine | remarkable |
| skyrocket | be subjected to | consist of | on behalf of | | |

1. It is difficult to _____ legend from truth.
2. Her team _____ children from Devon and Cornwall.
3. _____ changes have taken place in the jade market.
4. Production has dropped while prices and unemployment have _____.
5. The agent spoke _____ his principal.
6. We _____ the right to lodge a claim for loss.

7. I'm assured by the guarantee that the diamond is _____.
8. Details of the scheme have not yet been _____ to the public.
9. Innocent civilians are being arrested and _____ inhumane treatment.
10. The house is more than twenty meters long and _____ six meters in width.

Ⅲ. Grammar Work

Observe the following sentences and pay special attention to the use of attributive clauses.

1. The collections *we obtained recently* were amassed with great care over many generations.
2. There is no red reaction, *which suggests there is no color treatment*.
3. The bangle measures 6.5 *which is the hardness of jadeite*.

Now correct the mistakes in the following sentences.

1. This is the longest train which I have ever seen.
2. Which we all know, swimming is a very good sport.
3. Chapin, for who money was no problem, started a new film company with his friends.
4. His parents wouldn't let him marry anyone whom family was poor.
5. Emma became the first American woman to win three Olympic gold medals in track, that made her mother very proud.
6. The largest collection ever found in England was one of about 200,000 silver pennies, all of which over 600 years old.
7. The room that window faces the east is my bedroom.
8. She heard a terrible noise, as brought her heart into her mouth.
9. The day will come which the people all over the world will win liberation.
10. He paid the boy ten dollars for washing ten windows, most of them hadn't been cleaned for at least ten months.

Ⅳ. Translation

1. When subjected to the ultraviolet radiation the bangle does not release any ultraviolet light, which is a sign of natural grade A jadeite.
2. The absorption lines are in the areas representative of natural grade A jadeite.
3. To conclusively test for dyed or treated jadeite for the most sophisticated treatments, highly expensive equipment is needed, such as a Raman spectrometer and a Fourier transform infrared spectrometer.
4. Nephrite first appeared in China about 5000 BC but jadeite was not documented to appear in China until the mid 18th century when large deposits were imported into China.
5. 传统的翡翠鉴定方法主要依靠折射仪、宝石显微镜、查尔斯滤色镜和分光镜。
6. 宝石级翡翠为多晶质集合体的岩石,是著名的两种玉石材料之一。
7. 翡翠集高强的韧性和精美漂亮于一体,是理想的雕刻、珠串加工材料。
8. 软玉是不可再生资源,上好的和田玉原料更是非常稀有。

Part III Extended Learning

Dictation

Listen to the audio and complete the following passage with the words or expressions you hear.

Geologically speaking, jade is just one type of stone in the Earth's great family of stones. Besides the 1. _____ in China, there are other countries 2. _____. There are reserves of jade in the United States and Canada. There are reserves in Russia and South Korea. There are jade deposits in Poland, Italy, Zimbabwe and the New Zealand.

The fact is jade is just another type of natural stone, and is also found in many other parts of the world. 3. _____, it's hard to explain why this 4. _____ and long lasting jade culture should ever have emerged in China. And this raises an obvious question: what was it about jade that encouraged the ancient Chinese to plant the seeds from which this uniquely rich jade culture would have grown.

The earliest Chinese jade objects were 5. _____ at Xinglongwa, an age-old village in a city of China's Inner Mongolia Autonomous Region. At the site, archeologists discovered more than ten 6. _____ crafted more than 8,000 years ago. Scholars explained the birth of these primitive jade objects as follows:

At the beginning of human civilization, the most primitive aesthetic awareness in the minds of the ancients revealed itself and improved 7. _____. The ancients collected various interesting stones, shells and animal teeth and made them into simple ornaments, thus initiating 8. _____. Over time, natural jade gradually won the favor of the ancients due to its beautiful color and 9. _____.

Undoubtedly, it is the beautiful appearance of jade that makes it so easily distinguishable from other types of stone, but 10. _____. Jade is not only harder than common stone, it is tougher than most metals.

Read More

Passage 1

Chinese Jade Culture

Jade (*Yu* in Chinese Pinyin) was defined as beautiful stones by Xu Shen (about 58 – 147) in *Shuo Wen Jie Zi*, the first Chinese dictionary. Jade is generally classified into soft jade (nephrite) and hard jade (jadeite). Since China only had the soft jade until jadeite was imported from Burma during the Qing dynasty (1271 – 1368), jade traditionally refers to the soft jade so it is also called traditional jade. Jadeite is called *Feicui* in Chinese. *Feicui* is now more popular and valuable than the soft jade in China.

The history of jade is as long as the Chinese civilization. Archaeologists have found jade objects from the early Neolithic period (about 7000 BC), represented by the Hemudu culture in Zhejiang Province, and from the middle and late Neolithic period, represented by the Hongshan culture along the West Liao River, the Longshan culture along the Yellow River, and the Liangzhu culture in the Tai Lake region. Jade has been ever more popular till today.

The Chinese love jade because of not only its beauty, but also more importantly its culture, meaning and humanity, as Confucius (551 – 479 BC) said there are 11 De (virtue) in jade. The following is the translation:

The wise have likened jade to virtue. For them, its polish and brilliancy represent the whole of purity; its perfect compactness and extreme hardness represent the sureness of intelligence; its angles, which do not cut, although they seem sharp, represent justice; the pure and prolonged sound, which it gives forth when one strikes it, represents music. Its color represents loyalty; its interior flaws, always showing themselves through the transparency, call to mind sincerity; its iridescent brightness represents heaven; its admirable substance, born of mountain and of water, represents the earth. Used alone without ornamentation it represents chastity. The price that the entire world attaches to it represents the truth. The *Book of Songs* says, "When I think of a wise man, his merits appear to be like jade."

Thus jade is really special in Chinese culture, also as the Chinese saying goes "Gold has a value; jade is invaluable."

Because jade stands for beauty, grace and purity, it has been used in many Chinese idioms or phrases to denote beautiful things or people, such as Yu Jie Bing Qing (pure and noble), Ting Ting Yu Li (fair, slim and graceful) and Yu Nü (beautiful girl). The Chinese character Yu is often used in Chinese names.

There are Chinese stories about jade. The two most famous stories are *He Shi Bi* and *Wan Bi Gui Zhao*. *Bi* also means jade. *He Shi Bi* is a story about the suffering of a person named Bian He when he presented his raw jade to the kings again and again. The raw jade was eventually recognized as an invaluable jade and was named after Bian He by Wenwang, the king of the Chu State (about 689 BC). *Wan Bi Gui Zhao* is a follow-up story of the famous jade. The king of the Qin State, the most powerful state during the Warring States Period (475 – 221 BC), tempted to exchange the jade from the Zhao State using his 15 cities, but he failed. The jade was returned to the Zhao State safely. Thus jade is not only invaluable, but also the symbol of power in the ancient time. And it is interesting to note that the Supreme Deity of Taoism has the name, Yu Huang Da Di (the Jade Emperor).

Jade was made into sacrificial vessel, tools, ornaments, utensils and many other items. There were ancient music instruments made out of jade, such as jade flute, Yuxiao (a vertical jade flute) and jade chime. Jade was also mysterious to the Chinese in the ancient time so jade wares were popular as sacrificial vessels and were often buried with the dead. To preserve the body of the dead, Liu Sheng, the ruler of the Zhongshan State (113 BC) was buried in the jade burial suit composed of 2,498 pieces of jade, sewn together with gold thread.

Jade culture is very rich in China. We have only touched the surface of it. In conclusion, jade symbolizes beauty, nobility, perfection, constancy, power, and immortality in Chinese culture.

Passage 2

Five Tips on Buying Jade Jewelry

Buying jade jewelry is increasingly popular with both talented amateurs and souvenir hunters. However, while it was once comparatively easy to pick up a cheap piece of jade jewelry for cheap in places like Hong Kong and Taiwan, times have long changed. Buying jade for the right price, let alone a good price, is increasingly difficult.

Hong Kong and Taiwan remain the center for the jade trade but the market is now flooded with imitation jade—sometimes being passed off as genuine—and lower grade jade. There are very few genuine bargains available. The jade in Hong Kong and Taiwan is still competitively priced and you can save money on jade here, just not the legendary savings of yesteryear. Ultimately, real jade is expensive.

If you're in Hong Kong, your main port of call will be the Jade Market in Tsim Sha Tsiu, where dozens of sellers are clustered. The vast majority of the stuff on the sprawl of stalls are actually trinkets and souvenirs but there are also serious sellers here doing business.

1. Do your research

As with anything you buy—especially if it's a large investment—make sure you know what you want. There is a huge variety of jade, from yellowish and opaque stones to deep green, and different types have different price tags (deep green and white jade attract the highest prices and are relatively rare). Whichever type of jade you are interested in buying, be sure to check the prices at jewelers', both in your home country and wherever you're buying the jade. Shop around. This will give you an idea of the market value of certain pieces and allow you to make smart decisions about the price when negotiating.

2. Can't believe the price? Then don't

This is really a simple tip but it still seems to catch a lot of folks out. If a seller is offering a stone for well below market price, claiming it's pure, deep green jade and/or thousand years old, just walk away. These sorts of bargains don't exist. Scams on the other hand do exist.

3. Check the stone

We don't purport to be jade experts and if you are making any sort of major investment in jade, you certainly need to get the piece checked out by an independent expert. That said, there are some tips that can help you spot real jade. When buying jade the stone should be smooth and cool to the touch. Real jade is also very tough, so there should be no scratches on the surface. If you can scratch the stone with your fingernail, it's an imitation. Fake jade can often feel lighter than the real thing.

4. Get a certificate

For a big ticket piece of jade, the seller should be able to offer you a certificate of authenticity. This is jade that has been scanned and checked. The classification runs from A – D and is used internationally, although most sellers will only hold certificates for Grade A jade. It's worth noting that all grades—from A – D—are "real jade" just of differing quality and value. In grades B and C the jade has usually been treated with chemicals and may have had color added.

5. Imitation jade isn't always the enemy

Glass, plastic and all other sorts of minerals are used to create imitation jade, while the lowest opaque jade can be relatively inexpensive. As long as the seller is upfront about exactly what they are selling, there is no reason you shouldn't pick up a piece of imitation jade. Some of the pieces are very attractive.

Group Project

1. Choose one of the other polycrystalline gemstones and write a passage about it, giving as many details as possible.
2. Compare the differences between the Chinese people and Westerners in viewing and treating jade, and give a presentation on it.
3. Study the current development in the jade market and write a report about it.

Words and Expressions

polycrystalline /ˌpɒlɪˈkrɪstəlaɪn/	adj.	多晶质的
jadeite /ˈdʒeɪdaɪt/	n.	翡翠
nephrite /ˈnefraɪt/	n.	软玉（和田玉）
turquoise /ˈtɜːkwɔɪz/	n.	绿松石
serpentine /ˈsɜːp(ə)ntaɪn/	adj.	蜿蜒的
	n.	蛇纹石
quartzite /ˈkwɔːtsaɪt/	n.	石英岩玉
species /ˈspiːʃiːz/	n.	品种
agate /ˈægət/	n.	玛瑙
imitation /ˌɪmɪˈteɪʃ(ə)n/	n.	仿造；仿制品
deposit /dɪˈpɒzɪt/	n.	矿床
ornament /ˈɔːnəm(ə)nt/	n.	饰品
carving /ˈkɑːvɪŋ/	n.	雕刻，雕刻品
bracelet /ˈbreɪslət/	n.	手镯，手链

bangle /ˈbæŋg(ə)l/	n.	手镯，脚镯
transparency /trænˈspærənsi/	n.	透明度
translucent /trænsˈluːs(ə)nt/	adj.	半透明的
opaque /ə(ʊ)ˈpeɪk/	adj.	不透明的
dimension /daɪˈmenʃ(ə)n/	n.	尺寸；容积；维度
appraisal /əˈpreɪz(ə)l/	n.	鉴定
cylindrical /sɪˈlɪndrɪk(ə)l/	adj.	圆柱形的
consistency /kənˈsɪst(ə)nsi/	n.	连贯；浓稠度；硬度
genuine /ˈdʒenjʊɪn/	adj.	真正的，真实的；真诚的
provenance /ˈprɒv(ə)nəns/	n.	出处，起源
reserve /rɪˈzɜːv/	vt.	保留，保存
	n.	储备，贮存；预留品
specialist /ˈspeʃ(ə)lɪst/	n.	专家
polariscope /pəˈlærɪskəʊp/	n.	偏光镜
subject /sʌbˈdʒekt/	vt.	使……隶属
	adj.	须服从……的
radiation /reɪdɪˈeɪʃ(ə)n/	n.	放射，辐射；放射物
release /rɪˈliːs/	n.	释放，排放，解除
	vt.	释放；放开；发布
magnification /ˌmæɡnɪfɪˈkeɪʃ(ə)n/	n.	放大，夸大；放大率
hardness /ˈhɑːdnəs/	n.	硬度；坚硬；困难
measure /ˈmeʒə/	v.	测量；估量
	n.	测量，测度；措施；程度
spectrometer /spekˈtrɒmɪtə/	n.	分光仪，分光镜
craftsmanship /ˈkrɑːftsmənʃɪp/	n.	技艺，技术
polish /ˈpɒlɪʃ/	v.	擦光；磨光；润色
	n.	擦亮，磨光；优美，精良
sophisticated /səˈfɪstɪkeɪtɪd/	adj.	精致的；富有经验的；深奥微妙的
guarantee /ɡærənˈtiː/	n.	保证，担保；保证人，保证书
	vt.	保证，担保
eligible /ˈelɪdʒɪb(ə)l/	adj.	合适的；合格的；有资格当选的；称心如意的
antique /ænˈtiːk/	adj.	古老的；古董的
	n.	古玩，古董；古风
fracture /ˈfræktʃə/	vt.	（使）折断，破碎
	n.	破裂，断裂；骨折
diameter /daɪˈæmɪtə/	n.	直径，直径长；放大率
remarkable /rɪˈmɑːkəb(ə)l/	adj.	异常的，引人注目的；卓越的；显著的；非凡的

origin /ˈɒrɪdʒɪn/	n.	起源，根源；出身
skyrocket /ˈskaɪˌrɒkət/	vi.	突升，猛涨
Confucius /kənˈfjuːʃəs/	n.	孔子
virtue /ˈvɜːtʃuː/	n.	美德；德行；价值；长处
compactness /kəmˈpæktnəs/	n.	紧密；坚实；紧凑；小巧
iridescent /ˌɪrɪˈdes(ə)nt/	adj.	彩虹色的；变色的；闪色的；
ornamentation /ˌɔːnəmenˈteɪʃ(ə)n/	n.	装饰，装饰品
chastity /ˈtʃæstɪti/	n.	纯洁，贞洁，纯朴；贞操；童贞
denote /dɪˈnəʊt/	vt.	指代；预示；代表；意思是
invaluable /ɪnˈvæljʊ(ə)b(ə)l/	adj.	非常宝贵的；无法估计的；无价的
tempt /tempt/	vt.	引诱，怂恿；吸引；冒……的风险； 使感兴趣
symbolize /ˈsɪmbəlaɪz/	vt.	象征；用符号表现
sacrificial /ˌsækrəˈfɪʃ(ə)l/	adj.	牺牲的，祭祀的
utensil /juːˈtens(ə)l/	n.	器具，用具；器皿
immortality /ˌɪmɔːˈtælɪti/	n.	不朽，不朽的声名
amateur /ˈæmətʃə/	n.	业余爱好者；外行，生手
cluster /ˈklʌstə/	vt.	使密集，使聚集
sprawl /sprɔːl/	vi.	伸开四肢；蔓延
	n.	随意扩展；蔓延
stall /stɔːl/	n.	货摊
scam /skæm/	n.	骗局；诡计
purport /pəˈpɔːt/	vt.	声称；意图；意味着；打算
scratch /skrætʃ/	vt.	擦，刮；擦伤
	n.	擦，刮；刮擦声；搔痕
chime /tʃaɪm/	n.	合奏钟声，钟乐；谐音，韵律
preserve /prɪˈzɜːv/	vt.	保护；保持，保存
nobility /nəʊˈbɪlɪti/	n.	贵族阶级；高贵的身份；高尚，崇高；庄严，雄伟
thermal conductivity		导热性，热传导性
Chelsea filter		查尔斯滤色镜
color treatment		颜色处理
ultraviolet radiation		紫外线照射，紫外线辐射
ultraviolet light		紫外光
Raman spectrometer		拉曼光谱仪
reserve the right to…		保留……的权利
of high quality		高质量的
on behalf of		代表
Neolithic Period		新石器时期

music instrument	乐器
vertical jade flute	玉竖笛
pick up	购买
be flooded with	充满，充斥
pass sth. off as…	把……冒充为
yesteryear	去年，过往
certificate of authenticity	防伪证书
It's worth noting that…	值得注意的是……
there is no reason…	没有理由，没有道理

Unit 8

Pearl

Starting Out

☞ Match Words with Pictures

Match the words with the corresponding pictures.

| black pearl   round pearl   shell   teardrop pearl   oyster   baroque pearl |

1. _____   2. _____   3. _____

4. _____   5. _____   6. _____

☞ Check Your Knowledge

Fill the form with Chinese according to the given English.

English	中文	English	中文
luster		freshwater pearl	
saltwater pearl		natural pearl	
cultured pearl		round pearl	
teardrop pearl		pendant	
mollusk		pearl oyster	
nacre		grade	

Part I Communicative Activities

Order and Delivery

Task 1 Conversations

Ⅰ. **Read the following conversation and underline some useful sentences for buying and selling jewelry.**

(S: jewelry seller; C: customer)

S: I wonder if you have found that our specification meets your requirements. I'm sure the prices submitted are competitive.

C: Oh, yes, and I've come to place an order with you. We like the design of your jewelry.

S: My company will send you an official confirmation soon. But there are a few questions sought to be settled, such as the cost for sending the goods.

C: Yes, I see.

S: We quoted you as warehouse price if you want me to give you the price FOB. That would cover the transport from our house to deck and all the handling and shipping charges, leaving to pay the sea freights and marine insurance. Is that what you want?

C: No. I think we should prefer to have an idea of the total costs delivering right to our port.

S: Then what about a CIF price? That would cover the cost of the goods, a comprehensive insurance with a clause from warehouse to warehouse. All the forwarding and shipping charges are paid.

C: But there will be a few things left for us to pay.

S: Yes, the charges for your forward agent for clearing the goods, paying custom duties and arranging delivery to your sight. I can get the CIF prices worked out when we go on talking.

Ⅱ. **Look at the following conversation. Decide where the following sentences go and then act out the conversation in pairs.**

a. It will be a great honor for us to serve your business before we could move forward with the transaction.
b. That sounds a good idea.
c. Much to our regret, Mr. Mater.
d. We are very interested for the time-being in importing some jewelry of your company's design.
e. It's always being my great pleasure to meet and talk to my old friend.
f. I have been informed of the matter.

English for Jewelry

(S: jewelry seller; C: customer)

S: Good morning, Mr. Mater, welcome to our factory again.

C: Good morning, Mr. Chen. 1. _____

S: Likewise, how can I help you this time then?

C: Yes. As you know, we are one of the leading manufacturers of jewelry products in India. 2. _____ It will be highly appreciated if you could give us your firm's favorable quotation for the item.

S: 3. _____ However, we may have to explain that we could hardly supply you with the products directly at present since we are confined by the exclusive agent agreement with another company in India.

C: 4. _____ What I am talking about is the business between you and our established trade company in Hong Kong that channels your products into India and eludes the exclusive agent problem at the same time.

S: 5. _____

C: I'd like to book an order for 20,000 pieces in one lot.

S: 6. _____ I'm afraid we are not in the position to satisfy you with the quantity you required, as Spring Festival is usually the season for jewelry production. What's more, two of our workshops will be overhauled during the period and that will make the shortage even more strengthened.

Task 2 Role-play

Work in pairs and act out the following roles in the conversation about ordering and delivering jewelry.

Student A: a jewelry seller
Student B: a customer

Useful Expressions and Sentences

1. **Ordering jewelry**

- I've come to place an order with you. 我是来跟你下订单的。
- We are very interested for the time-being in importing… 目前我们想进口……
- It will be highly appreciated if you could give us your firm's favorable quotation for the item. 如果你们能提供货物的优惠报价，那就太好了。
- I'd like to book an order for…pieces in one lot. 我想预订一批……件的订单。
- I'm afraid we are not in the position to satisfy you with the quantity you required, as…is usually the season for jewelry production. 非常抱歉，我们没法满足你要求的数量，因为……通常是珠宝的旺季。
- I wonder if you have found that our specification meets your requirements. I'm sure the prices submitted are competitive. 不知道我们的规格是否满足你的要求。我相信我们提供的价格是极具竞争力的。

2. Delivering jewelry

- My company will send you an official confirmation soon. But there are a few questions sought to be settled, such as the cost for sending the goods. 我们公司会尽快给你一份正式的确认书。不过还有几个问题要确定下来，例如运输费用。
- We quoted you as warehouse price if you want me to give you the price FOB. That would cover the transport from our house to deck and all the handling and shipping charges, leaving to pay the sea freights and marine insurance. 如果你想以 FOB（离岸价格）结算，我们就报货仓价格，包括从我们仓库到甲板的运输费和包装费。你只需要另付海运费和海运保险。
- I think we should prefer to have an idea of the total costs delivering right to our port. 我们更想知道货物到我们港口的总运输费用。
- Then what about a CIF price? That would cover the cost of the goods, a comprehensive insurance with a clause from warehouse to warehouse. 那以 CIF（到岸价格）结算如何？包括货品价格、全面的保险以及从我们仓库到你们仓库的运输费。
- I can get the CIF prices worked out when we go on talking. 协商过程中，我就可以计算出 CIF 的总价。

Part II   Read and Explore

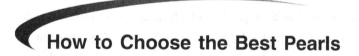

How to Choose the Best Pearls

Pearls are very beautiful and popular in the jewelry world. Although they are present in many pieces of jewelry, it can often be hard to know how to choose the best pearls. If you are planning on purchasing some pearls, you should familiarize yourself with them prior to purchase. Learning a little about their origin and characteristics will allow you to choose the best pearls.

The Origin of Pearls

In order to choose the best pearls, you should become a bit familiar with them. First of all, a pearl is formed within the body of an oyster or other small mollusk. This formation takes place when an irritant of some sort enters the mollusk's body or shell. Whether this irritant is a parasite, a grain of sand, or something else, the mollusk reacts by creating a hard, protective layer of material around it. This protective material is known as nacre and is made up of calcium carbonate. The nacre will continue to grow, layer by layer, until a pearl is formed. Pearls can be found in both freshwater and saltwater mollusks.

Types of Pearls

There are two main types of pearls available. Most pearls on the market today are cultured pearls. Cultured pearls are created and tended to by humans. To form cultured pearls, humans actually place an irritant inside the shell of a mollusk to instigate the pearl growing process. Once the irritant is introduced, the mollusk begins to grow layers of calcium carbonate around it. Natural pearls are the other main type of pearls. They are formed randomly in nature and are much less common than cultured pearls. These pearls are rare because an irritant must lodge itself within the mollusk, instead of a person doing so. Because natural pearls are uncommon, they are much more expensive than cultured pearls. Now that you know a little about pearls, you can learn how to choose the best.

Choosing the Best Pearls

The first thing you want to look for when choosing pearls is their luster. Luster is the sheen and iridescence of the pearls' surface and is determined by how the layers of nacre absorb and reflect light. The better the sheen and iridescence of the pearls, the more high quality they are. Color is another factor to take into consideration when choosing pearls. There are light pearls, which generally come in tones of white and ivory. Light colored pearls can also be found with rose or silver overtones, making them more desirable 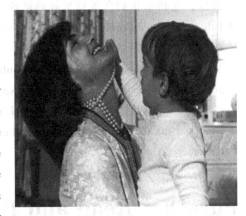 to many. There are also dark pearls, which are usually black with green or blue overtones. Dark pearls are much rarer than light pearls, which makes them more valuable as well. You should know that pearls can actually be dyed to many colors too. The shape of pearls is also important. The best pearls will be perfectly symmetrical with a smooth surface that is free of imperfections. Symmetrical pearls can be found in various shapes including round, oval, or tear drop. The size of pearls can also make a difference in their value. Larger pearls are obviously harder to grow, making them more scarce and valuable. It should also be known that pearls are graded on their quality and appearance. Because the grading system can vary greatly, it is best to ask your jeweler to find out the details.

The Bottom Line

To choose the best pearls, first examine their luster and ask about their grade. These are the main factors that must be taken into consideration if you want the best pearls. Next, make sure all the pearls on the piece of jewelry are symmetrical and similar in appearance. If you choose colored pearls, be sure to find out their origin. Your budget will also greatly impact the pearls you choose. Cultured pearls are much more common than natural pearls, making them more affordable.

If price is no object, you can choose the best natural pearls you can afford.

Now that you know some general information on them, you should be able to choose the best pearls. Just remember to shop around at various jewelers before making a decision. This will allow you to see many different choices of pearls available.

Check Your Understanding

I. Fill in the form about the description of different aspects on choosing pearls.

Aspects		Description
luster		
color	light pearls	
	dark pearls	
shape		
size		

II. Answer the following questions.

1. How are pearls formed?
2. How many types of pearls does the author introduce?
3. What is the difference between luster and color in choosing pearls?
4. What is "grade"?
5. How does budget influence your choosing pearls?

Subject Focus

1. Talk about the formation of pearls. Then have a check about whether you've made the description right and clear with your partner.
2. Talk about types of pearls you know. Then exchange opinions with your classmates.
3. Suppose you are a fashion designer, design a piece of jewelry with pearls and make a draft. Then present the jewelry to the class with your oral explanation.

Language Focus

I. Subject-related Terms

Fill in the blanks with the words or expressions being defined.

1. _____ a sum of money allocated for a particular purpose
2. _____ assign a rank or rating to
3. _____ single thickness of some homogeneous substance
4. _____ exhibiting equivalence or correspondence among constituents of an entity
5. _____ the hard largely calcareous covering of a mollusk
6. _____ the iridescent internal layer of a mollusk shell
7. _____ a smooth and gentle brightness

8. _____ the visual property of something that shines with reflected light
9. _____ something that causes irritation and annoyance
10. _____ invertebrate having a soft body usually enclosed in a shell

II. Working with Words and Expressions

In the box below are some of the words and expressions you have learned in this text. Complete the following sentences with them. Change the form of words if necessary.

familiarize...with...	prior to	take place	randomly	look for
take into consideration	determine	various	reflect	be free of

1. How many of you believe that there are meetings that _____ without you?
2. He began to _____ a new job immediately after he was fired.
3. The size of the chicken pieces will _____ the cooking time.
4. His plan is to spread the capital between _____ building society accounts.
5. I viewed it as important, but as something which came after, not _____ activity.
6. Their actions clearly _____ their thoughts.
7. It mimics a memory leak by _____ placing some of these arrays into a list.
8. Of course, we also have to _____ our lifestyles.
9. The goal of the experiment was to _____ the people _____ the new laws.
10. These models must _____ errors before you import them.

III. Grammar Work

Observe the following sentences and pay special attention to the use of preposition.
1. Pearls are very beautiful and popular *in* the jewelry world.
2. First of all, a pearl is formed *within* the body of an oyster or other small mollusk.
3. Color is another factor to take *into* consideration when choosing pearls.

Now correct the mistakes in the following sentences.
1. We often play football in Saturday afternoon.
2. You are wanted by the phone.
3. This is an introduction in our products.
4. As he was sick, he asked about a leave of absence.
5. She married to a man she didn't love at all.
6. I will call on your office tomorrow.
7. He told me he didn't agree with the plan.
8. It's kind for you to come to see us.
9. He makes a living for teaching.
10. They said they'd be in holiday in the countryside.

IV. Translation

1. If you are planning on purchasing some pearls, you should familiarize yourself with them prior

to purchase.
2. This formation takes place when an irritant of some sort enters the mollusk's body or shell.
3. Whether this irritant is a parasite, a grain of sand, or something else, the mollusk reacts by creating a hard, protective layer of material around it.
4. Luster is the sheen and iridescence of the pearls' surface and is determined by how the layers of nacre absorb and reflect light.
5. 市面上的珍珠主要分为两大类：养殖珍珠和天然珍珠。
6. 根据不同的质量和外观，珍珠会被评级。
7. 挑选最好的珍珠，首先检查它们的色泽，并询问它们的评级。
8. 浅色珍珠还暗含粉红色和银白色，是许多人喜欢的颜色。

Part Ⅲ Extended Learning

 Dictation

Listen to the audio and complete the following passage with the words or expressions you hear.

How to Do a Pearl Appraisal

Hi. I'm Abbott Taylor, from Abbott Taylor Jewelers in Tucson, Arizona, here for About. com.

Now we're going to talk about pearls and the different qualities that appraisers look at to give them their value. Pearls have major factors that are used to describe the quality of a pearl. 1. _____ and it's approach from regular to round. The more it gets to round, the more expensive it's going to be. Then there's 2. _____. Luster has to do with the surface reflectivity. If it glows and you see the change in light in the room around you and you might almost be able to see a reflection in the surface of the pearl.

Surface Blemishes

You can have a perfectly round pearl or a 3. _____ pearl or a very irregular 4. _____ pearl and the surfaces can be very smooth. If you have a round pearl that is perfectly smooth, that's most valuable.

If it has pock marks or dimples in it, or irregularities, or places where there's indentations, or abrasions or cut lines that go around the pearl, these are all things that reduce its value. Thickness of 5. _____ has to do with how thick the outer shell of the pearl is.

Cultured Pearls

Pearls are 6. _____ pearls. There's very little evidence that there are any live natural pearls in the wilderness left today. The only natural pearls, pearls that were not cultured, but found in nature, have long since left the planet. And they're all antiques—very, very valuable, mind you.

7. _____ pearls are a different story and they have to be checked by x-ray. If they're not 8. _____ then you can't look through the hole. Then you have to x-ray them to see whether they are natural or they are cultured.

Size and Color

Size: one of the more important factors is physically how large it is. Because the way cultured pearls are made, they take either a mother of pearl bead, made from the shell of an 9. _____, and they implant it into the muscle. They make it round, they make it perfectly shaped, and they implant it into the muscle of the oyster.

The oyster then produces the nacre that goes over the surface of that. And the longer it grows in that muscle the thicker the nacre gets. So size is very important. And it gets exponentially more expensive per half 10. _____ after it reaches eight millimeters.

The hue, and tone and color of the pearl: so, there's cream, white, and pink or rosé, the rosé being a much more valuable pearl than the white, and the white being more valuable than a cream.

 Read More

Passage 1

A Beginner's Guide to Pearls

Pearls are an organic gem, called organic because they are created by living creatures. Each pearl begins its existence as a piece of grit or other particle that makes its way into the shell of a marine or freshwater mollusk — some types of oysters and clams. A defense mechanism kicks-in and coats the particle with layer after layer of a substance called nacre, or mother-of-pearl, which eventually becomes thick enough to form a pearl.

Pearls are classified by their origins and their shapes.

Pearl Origin Classifications

Natural Pearls

Natural pearls are formed when an accidental intruder enters a mollusk's shell and continuous layers of nacre grow like onion skins around the particle. Natural pearls vary in shape depending on the shape of the piece being coated.

Natural pearls have always been considered rare and are quite expensive. They are usually sold by carat weight. The most natural pearls on today's market are vintage pearls.

Cultured Pearls

Like natural pearls, cultured pearls grow inside of a mollusk, but with human intervention. A shell is carefully opened and an object is inserted. Shapes of objects vary, depending on the final shape of pearl that's desired.

Over time the object becomes coated with layers of nacre. The depth of the nacre coating depends on the type of mollusk involved, the water it lives in, and how long the intruder is left in place before being harvested. As nacre thickness increases, so does the quality and durability of the cultured pearl. Cultured pearls are sold by their size in millimeters.

Saltwater Pearls

Saltwater pearls originate within a saltwater mollusk. Saltwater pearls can be either natural or cultured.

Freshwater Pearls

Freshwater pearls grow inside of a freshwater mollusk — one that lives in a river or a lake.

Pearl Shape Classifications

Spherical pearls are round, which is traditionally the most desirable shape. The rounder the pearl, the more expensive its price tag.

Symmetrical pearls include pear shaped pearls and other shapes that have symmetry from one side to another, but are not round.

Baroque pearls are irregularly shaped pearls. They are often the least expensive category of pearls, but are unique and quite beautiful.

Passage 2

How to Clean Pearls

The pearls most of us wear today are cultured pearls, their existence initiated by humans who insert a bead or other object into an oyster or clam. The clam coats the foreigner with nacre, the patina that gives pearls their unique appearance.

Caring for Your Pearls

Even cultured pearls with thicker coatings are more fragile than most other gemstones, so you must handle them carefully to keep them in the best condition.

Your pearls will stay cleaner if you put them on after you've applied your makeup and perfume.

Be sure to take off your pearl rings before you apply hand and body creams.

Wipe your pearls with a soft, lint-free cloth as soon as you take them off. The cloth can be dampened with water or it can be dry. If damp, allow the pearls to air dry before putting them away.

Dirty pearls can be cleaned with a mild soap and water solution.

Never clean your pearls with solutions that contain ammonia or harsh detergents.

Don't put pearl jewelry in an ultrasonic cleaner.

Don't use abrasive cleaners or rub pearls with abrasive cloth. Both can wear away the nacre coating, leaving you with a plain loo-

king bead.

Storing Pearls

Don't store your pearls with other jewelry, because they can be scratched easily when metal or gemstones rub against them. Find a special slot in your jewelry box for the pearls, or keep them in a soft bag made from chamois or another non-abrasive material.

Your fine pearl necklaces should be restrung periodically so that you're sure the silk or nylon cord holding them is in good shape.

Group Project

1. Choose one type of pearls and write a passage about its formation, giving as many details as possible.
2. Design a piece of jewelry with the type of pearl you've talked about above. You are allowed to use other materials to accomplish that jewelry.
3. Study the current fashion trend of pearl jewelry design and write a report about it.

Words and Expressions

organic /ɔːˈgænɪk/	adj.	有机的
creature /ˈkriːtʃə/	n.	生物
grit /grɪt/	n.	粗砂
particle /ˈpɑːtɪk(ə)l/	n.	颗粒
marine /məˈriːn/	adj.	海洋的
mollusk /ˈmɒləsk/	n.	软体动物
oyster /ˈɔɪstə/	n.	牡蛎
clam /klæm/	n.	蛤蜊
coat /kəʊt/	vt.	覆盖
layer /ˈleɪə/	n.	层，层次
substance /ˈsʌbst(ə)ns/	n.	物质
intruder /ɪnˈtruːdə/	n.	侵入者
insert /ɪnˈsɜːt/	vt.	插入，嵌入
harvest /ˈhɑːvɪst/	v.	收割
	n.	收获
durability /ˌdjʊərəˈbɪləti/	n.	耐久性
millimeter /ˈmɪlɪˌmiːtə/	n.	毫米
initiate /ɪˈnɪʃɪeɪt/	v.	开始
patina /ˈpætɪnə/	n.	铜绿
fragile /ˈfrædʒaɪl/	adj.	脆弱的

make-up /ˈmeɪkˌʌp/	n.	化妆品
dampen /ˈdæmp(ə)n/	vt.	使……潮湿
ammonia /əˈməʊnɪə/	n.	氨
rub /rʌb/	v.	擦
scratch /skrætʃ/	vt.	刮
slot /slɒt/	n.	狭槽

defense mechanism	防御机制
mother-of-pearl	珍珠母
carat weight	克拉重量
vintage pearl	古珍珠
spherical pearl	球形珍珠
symmetrical pearl	对称珍珠
baroque pearl	巴洛克珍珠
air dry	风干
harsh detergent	刺激性洗涤剂
ultrasonic cleaner	超声清洗器
abrasive cleaner	磨砂清洁器
nylon cord	尼龙绳

Unit 9

Gemstone Design

Starting Out

☞ **Match Words with Pictures**

Match the words with the corresponding pictures.

| flower round cut ruby | natural top green emerald oval | twinkle oval amethyst |
| blue sapphire octagon shape | round brilliant zircon | blue zircon |

1. _____

2. _____

3. _____

4. _____

5. _____

6. _____

☞ **Check Your Knowledge**

Fill in the English expressions of the following Chinese terms with teacher's help.

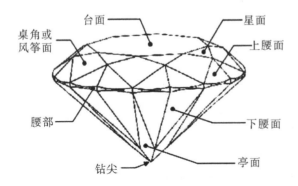

1. 台面 _____
2. 星面 _____
3. 上腰面 _____
4. 下腰面 _____
5. 亭面 _____
6. 钻尖 _____
7. 腰部 _____
8. 桌角或风筝面 _____

Part I Communicative Activities

Payment Terms

Task 1 Conversations

Ⅰ. Read the following conversation and underline some useful sentences for buying and selling jewelry.

(S: sales assistant; B: buyer)

S: Would you like jewelry?
B: Yes, of course.
S: What kind of jewelry do you like to have?
B: I'd like 14K and 18K gold necklaces, chains and earrings.
S: Sure. Here is a nice gold necklace.
B: Can I have a look?
S: Here you are. Its regular price is $5,600 and now you can have it with a ten percent discount.
B: It's very elegant. I'll take it. Besides, I should like to look at some bracelets.
S: Jadeite or nephrite?
B: I don't know what you said, but I love the bracelets with light purple.
S: Oh, that's jadeite from Burma.
B: What's the price for it?
S: $1,500.
B: How about $900?
S: I'm sorry we only sell at fixed prices.
B: It's a pity. I'm ready to give it up. Food is more important than jewelry, I think.
S: Ha-ha. I see. This is the bill of the necklace.
B: OK. Which do you prefer, cash or credit card?
S: Both are OK. The counter is on the right.
B: OK. I'm going to pay the check.

Ⅱ. Look at the following conversation. Decide where the following sentences go and then act out the conversation in pairs.

> a. Let me try it on.
> b. I'll pay the messenger on delivery.
> c. I'll give you a check for the amount on the Bank of China.
> d. How many carats would you like it to be?
> e. our shop doesn't ask two prices.

(A: sales assistant; B: buyer)

S: Welcome to our shop. Can I help you?
B: I wish to buy a diamond ring.
S: 1. _____
B: I want one carat.
S: Is this one suitable for you?
B: No, it seems too old-fashioned to me.
S: How about this?
B: 2. _____ It's too small for me. Haven't you got any larger ones?
S: Then you may take that one. It's very nice and the latest in style.
B: This fits me well. How much do you charge for it?
S: Eight thousand and two hundred dollars.
B: It's too expensive. I can only pay you seven thousand dollars.
S: I told you before, lady, 3. _____
B: Good, I'll have it. Have you got any brooches?
S: With diamond, ruby or sapphire?
B: Sapphire, please. How much is it?
S: Four hundred.
B: All right. How much will it be altogether? Please send it to my address, 4. _____
S: Sorry. It's our rule never to supply goods cash on delivery.
B: Well then, 5. _____

Task 2   Role-play

Work in pairs and act out the following roles in the conversation about payment terms.

Student A: a sales assistant in a jewelry store

Student B: a consumer buying a necklace and a brooch

Useful Expressions and Sentences

- A bill was received and duly paid. 收到了账单并及时支付了。
- The last consignment was found defective. 发现最后一批货物有瑕疵。
- We regret that we cannot comply with your request. 很遗憾，不能满足你方的要求。
- Please foot this bill on the counter; I'll pack the bracelet for you. 请付款，我给您包装这个手镯。
- The whole set is cheaper. We have offered a ten percent discount. We can't reduce the price any further. 整套买更便宜。我们已经按九折报价了，不能再降价了。

Part II Read and Explore

Production Process and Technology of Jewelry

Traditionally skilled goldsmiths employing craft skills and simple hand tools have made jewelry and even nowadays, worldwide, much is still made in basic workshops using manual metalworking skills and only limited use of machines. Increasingly, however, new mass production methods are being adopted incorporating precision engineering techniques like CAD (Computer Aided Design) and lasers capable of producing high quality jewelry indistinguishable from traditionally hand made pieces. Any outfit should be built around a core of skilled craftsmen using modern equipment.

Good design is essential, and is especially crucial and significant as a basis for manufacturing mass-produced jewelry, since any design mistakes are very costly, both in terms of labor and tooling costs. Most good designers are craftsmen jewelers, who understand the materials and their limitations, and appreciate the functional problems. Necklaces, for example, must fall right and be comfortable to wear; earrings must not be too heavy; brooches must not tear flesh; and rings must not snag on material, and so on. A good designer must be aware of fashion trends and be able to predict what will be in vogue, for up to 2 years ahead.

There are three basic methods of making jewelry: handmade, die stamping and jig assembly manufacture or casting, either whole or in components. Once a piece is made up, by whatever method, it is passed to a finisher for smoothing and polishing and finally, if required, to a setter to complete gem settings.

Jewelers do not restrict themselves to one particular method, and, depending on what is to be made, will use, for maximum efficiency, any of the three methods, in combination, to achieve desired quality within a price range.

Handmade jewelry

The raw material is first made up to the design by mounters using ready formed metals, in the form of sheets, wires, tubes and findings (ready made joints, catches and settings). The tools used are the traditional smith's: hammers, drills, punches, gravels, and files and a heat source, nowadays a blowtorch. With these, mounters fashion and assemble the pieces of jewelry, which are then passed on to the finisher and setter.

Machine production

The basic method of mechanized production is for components to be cut and stamped with individual steel dies, which are manufactured for each component of each individual item of jewelry. As die manufacturing is an expensive operation it is cost-effective only where high numbers of the same item are being produced. Once the individual components are made, they are fitted on jigs for the final assembly process, which can be completed by semi-skilled operators. In high end/

high volume operations, complex "progression" die sets are used in which a strip of work-in-progress passes in stages through a single die and emerges in the final shape.

Casting

The process of "lost wax" casting (using wax replicas of an original piece and also known as investment casting) in use since 4000BC lends itself easily to mass production, with the advantage of much lower capital costs. Sometimes, if making a complicated design or intricate piece of jewelry, it becomes necessary to solder up from a number of cast components. Even the solder needs to be in gold, silver or platinum alloy and must conform to assay requirements. Casting is resurging in popularity.

Electroforming

For most people, jewelry design brings to mind creativity and art. Concepts related to physics and technology seemingly have no place in the workshop of an artist. Yet the new technology of electroforming is opening up doors of opportunities to designers worldwide. It is worth the time of any serious jewelry designer to take a few moments to understand the processes behind this new technology.

What is electroforming? Electroforming is a highly specialized process of metal part fabrication using electrodeposition in a plating bath over a base form (called a mandrel, model or pattern) which is subsequently removed. Technically, it is a process of synthesizing a metal object by controlling the electrodeposition of metal passing through an electrolytic solution onto a metal or metalized form. More simply, a metal skin is built up on a metal surface, or any surface that has been rendered electroconductive through the application of a paint that contains metal particles. Essentially, a metal part is fabricated from the plating itself.

Why use electroforming? Compared with other basic metal forming processes (casting, forging, stamping, deep drawing, machining and fabricating), electroforming is very effective when requirements call for extreme tolerances, complexity or light weight. The precision and resolution inherent in the photographically produced conductive patterned substrate, allows finer geometries to be produced to tighter tolerances while maintaining superior edge definition with a near optical finish. Electroformed metal is extremely pure, with superior properties over wrought metal due to its refined crystal structure.

The advantages of electroforming are endless. Jewelry can be made in pure 24 karat gold as well as karat gold ranging from 8K to 18K. The electroforming process makes it possible to create thin, hollow items in complex, three-dimensional shapes. A variety of designs can be produced, including large modern shapes with low weight and uniform wall thickness, and there is no loss of metal in the process; no scrap is generated. The central disadvantage is that the electroforming technique requires special equipment designed for electroforming.

The technique of electroforming will undoubtedly have an impact on the jewelry industry. For thin, hollow, lightweight, voluminous, and complex three-dimensional shapes, electroforming is unparalleled. Electroforming allows for the realization of designs that are not possible by other techniques, at affordable prices. Among its many uses, electroforming can replicate natural objects—leaves, flowers, shells, and nuts, for example—as well as create interesting shapes from manmade objects. Classic applications for electroformed jewelry include earrings, pendants, brooches, chains and necklaces, charms, clasps, and bangles.

Check Your Understanding

I. **Fill in the blanks according to the text.**

1. Three basic methods of making jewelry mentioned in the passage are _____, _____ _____ or casting.
2. The basic method of mechanized production is for components to be _____ and _____ with individual steel dies.
3. For most people, jewelry design brings to mind _____ and _____.
4. Classic applications for electroformed jewelry include _____, _____, _____, chains and so on.
5. Among its many uses, electroforming can _____ and _____ objects.

II. **Answer the following questions.**

1. What's the use of good design in mass jewelry production?
2. What's the procedure of handmade jewelry production?
3. Under what condition is die manufacturing cost-effective?
4. What are the similarities between die manufacturing and casting?
5. Why use electroforming?

Subject Focus

1. Write a summary about different methods of jewelry production.
2. Make a market research on different methods of jewelry production and give a presentation about how to distinguish different methods of jewelry production.
3. Suppose you are gemstone store owner, and a market researcher is interviewing you about the popularity of gemstones made by different methods. Now make a conversation by using the knowledge in the text.

Language Focus

I. **Subject-related Terms**

Fill in the blanks with the words or expressions being defined.

1. _____ very small stones
2. _____ weighing less than most other things of the same type
3. _____ to invent to deceive others; to make out of other materials or substances
4. _____ small problem or disadvantage

English for Jewelry

5. _____ greater in intensity or more common
6. _____ inborn, innate
7. _____ difficult to know which is which or who is who
8. _____ extremely important
9. _____ accurate copy of something or somebody
10. _____ to produce something by means of chemical or biological reactions

II. Working with Words and Expressions

In the box below are some of the words and expressions you have learned in this text. Complete the following sentences with them. Change the form of words if necessary.

scrap	resurge	raw	file	replica
voluminous	pattern	subsequently	sterling	a strip of

1. The data can _____ be loaded on a computer for processing.
2. The Renaissance again discovered the world, the man and brought about the _____ of politics.
3. China and Japan are close neighbors separated only by _____ water.
4. Manicurists are skilled at shaping and _____ nails.
5. He went to prison for receiving stolen _____ iron.
6. I sank down into a _____ armchair.
7. Discharge of _____ sewage into the sea is unsanitary and unsafe.
8. Those are _____ qualities to be admired in anyone.
9. _____ weapons are indistinguishable from the real thing.
10. The bed linen is _____ in stylish checks, stripes, diagonals and triangles.

III. Grammar Work

Observe the following sentences and pay special attention to the use of Conjunction.

1. Good design is essential, and is especially crucial and significant as a basis for manufacturing mass-produced jewelry, *since* any design mistakes are very costly, both in terms of labor and tooling costs.
2. *As* die manufacturing is an expensive operation it is cost-effective only where high numbers of the same item are being produced.

Now correct the mistakes in the following sentences.

1. He failed the exam because of he was not well prepared.
2. As if he was not well, she had to go without him.
3. His request is unreasonable for that he knows we can't afford it.
4. They are hoping for a return to normality now since that the war is over.
5. As long as all of us try our utmost, the project will be finished on schedule.
6. The former hostage is in remarkably good shape in considering his ordeal.

7. Owing to the engine has trouble, the plane had to make a forced landing.
8. Thanks to the rain, we came back home all wet.
9. Seeing he was a hard worker, he achieved a lot.
10. He had a great desire to have a home of his own for that he had always lived with his grandmother.

IV. Translation

1. Most good designers are craftsmen jewelers, who understand the materials and their limitations, and appreciate the functional problems.
2. Once a piece is made up, by whatever method, it is passed to a finisher for smoothing and polishing and finally, if required, to a setter to complete gem settings.
3. For most people, jewelry design brings to mind creativity and art. Concepts related to physics and technology seemingly have no place in the workshop of an artist.
4. The central disadvantage is that the electroforming technique requires special equipment designed for electroforming.
5. 一个好的切工能改变展示宝石的颜色、减少夹杂物并呈现良好的整体对称性及切工比例。
6. 大多数不透明的宝石，包括欧泊、月光石等，一般都加工成弧面形而不加工成刻面。
7. 电铸在价格能承受的基础上实现了其他技术实现不了的设计。
8. 宝石匠不会局限于某一种方法；根据要做的产品，他们会最大效率地使用三种方法中的任何一种或结合使用，以便在某个价格范围内达到理想的质量。

Part III Extended Learning

 Dictation

Listen to the audio and complete the following passage with the words or expressions you hear.

Bollywood Actresses at Catwalk of India International Jewelry Week

The India International Jewelry Week has come to a close. Various designers showcased their jewelry 1. _____ on the final day, with a little help form some beautiful Bollywood actresses.

Actress Bipasha Basu strutted the catwalk in a green embroidered skirt and a maroon blouse, 2. _____ a gorgeous necklace, shoulder dusters, bangles and a cocktail ring.

Basu wore jewels created by Sachin and Nitin Gupta of PC Jeweler, and 3. _____ by Indian fashion designer Neeta Lulla. Models displayed "temple jewelry", featuring gold pieces with the embossed forms of great deities and gods.

Models also wore uncut diamond necklaces, with 4. _____ and dangling earrings.

"It's very elegant, very traditional, yet for the modern women. I think the kind of jewelry that I am wearing, I have chosen myself, so it's a 5. _____ of the modern woman's personality, so there is a good mix of modern and traditional," Indian actress Bipasha Basu said.

At a show of Indian jewelry brand Ganjam, models wore long bibs and chandelier earrings, medallions for necklaces, and 6. _____ carved bracelets.

They displayed designs in colored, blue and uncut diamonds, embellished with 7. _____ sparkling gems.

Bollywood actress Sonam Kapoor, 8. _____ for the India International Jewelry Week 2014, also walked for the finale. The highlight of her look was a bridal headband called "Arbor Amore", created in a gold leaf design.

"It's a statement, so I want the focus on this head piece. It's a lot of fun wearing this jungle jewelry, and with a sari, wearing 9. _____, I feel like a goddess, it is very nice," Sonam Kapoor, Indian actress said.

The four-day India International Jewelry Week aims to showcase the country's finest in jewelry, supported by top-of-the-line 10. _____, technology and quality, to customers around the world.

Read More

Passage 1

Gemstone Cut

Most of the gemstones are worn as waterworn and damaged or rounded pebbles, though some good crystals may also be obtained. The value and beauty of gemstones are very much enhanced by the proper cutting of facets, because the optical properties are then brought out to the best advantage. In cutting, it is always the aim to maintain the symmetry of the crystal.

The quality of a gemstone's cut can have a dramatic impact on how it looks but only a small impact on the price per carat. Jewelers and savvy gem shoppers are paying more attention to cut to get the most beauty for the money.

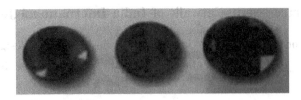

How can you tell if a gem is well cut? The easier way to train your eye is to look at bad cutting next to good cutting. Look at the picture above, which is of three rubies, all with good color and clarity. The ruby in the middle looks better because all the light that goes in is reflected back to the eye. See how the color is evenly distributed across the surface of the stone. This ruby also has more life and sparkle as the light dances across the facets. In contrast, the other two stones have dark areas where light is not reflected to the eye of the observer at the right angle.

Cutting style can also add beauty to a gemstone. Look at the above picture of two stones: they are the same size, shape, fine quality and color but dramatically different in appearance because one is a standard emerald cut and the other is a Barion cut, with more facets in the back.

In addition to these standard gemstone shapes, gem designers are inventing new ways of cutting gemstones in unique individual styles. For example, some facet gems in unusual geometrical shapes, some carve gemstones and some use a combination of faceting and carving.

A good cut showcases the gemstone's color, diminishes its inclusions, and exhibits good overall symmetry and proportion. Because gemstone color can vary, there are no hard geometrical standards when it comes to maximizing brilliance or color. Gemstones, especially rarer ones, are sometimes cut for size without regard for their color. For example, when corundum varieties such as sapphire and ruby are cut for maximum weight rather than beauty, they may display banded colors or streaks.

In a gemstone with more saturated color, the best cut may be more shallow than average, permitting more light to penetrate the gemstone, while in a less saturated gem, the color may benefit from a deeper cut.

Look at the gemstone in the setting and ensure that all the facets are symmetrical. An asymmetrically cut crown indicates a gemstone of low-quality. In all cases, a well-cut gemstone is symmetrical and reflects light evenly across the surface, and the polish is smooth, without any nicks or scratches. Like diamonds, fine quality color gems usually have a table, crown, girdle, pavilion, and culet. Iridescent opals are one exception, and most often have a rounded cabochon cut.

Most gems that are opaque rather than transparent are cut as cabochons rather than faceted, including opal, turquoise, onyx, moonstone. You will also see lower grade material in gemstones such as sapphire, ruby and garnet cut as cabs. If the gem material has very good color but is not sufficiently transparent or clean to be faceted, it can still be shaped and polished to be a very attractive cabochon. It is also common to cut softer stones as cabs, since gems with a hardness less than 7 (on the Mohs scale) can easily be scratched by the quartz in dust and grit. Minute scratches show much less on a cabochon than on a faceted stone.

Passage 2

Judging the Quality of the Setting

Now that you have found the perfect gem, all you need to do is make sure it is displayed well and held securely in place.

To judge the quality of the jewelry setting, pay close attention to details. Is the metal holding the stone even and smoothly finished so it won't catch on clothing? Is the stone held firmly and square in the setting? Is the metal well polished with no little burrs of metal or pockmarks?

Inexpensive jewelry is often very lightweight to give you a bigger look for the money. If a piece is lightweight, pay special attention to the prongs holding the stone: are they sturdy? Do they grip the stone tightly? You won't be happy about the money you saved in gold cost if you lose your stone!

If the piece is gold, does it have a karatage mark? Is the company trademark stamped next to it? If it is, the company is standing behind that mark and assuring you that the karatage is as stated.

When buying a necklace, make sure it lays well around the neck. Try it on or ask a salesclerk to model it so you can check how it fits against the skin. For earrings, check to make sure that they hang well from the ear and don't tip forward. Designs that are asymmetrical should have a left and right which mirror each other.

Here is one final hint from real jewelry buying pros: if a piece of jewelry is really well made, the back will be well finished also.

If you are buying a gift and you are not sure about the style of the piece of jewelry, why not give the perfect gemstone in a black velvet pouch and let the lucky recipient design her own perfect setting: many jewelers offer custom design services. Gems can speak louder than words. You can choose a gemstone that symbolizes what you want to say with your gift.

Group Project

1. Choose one kind of gemstones and write a passage about its cutting process, giving as many details as possible.
2. Present the process of jade cutting and processing, with pictures or even videos to help explain clearly.
3. Interview a jewelry processor and ask him/her to talk about some points to pay attention to in jewelry design.

Words and Expressions

metalworking /ˈmet(ə)lwɜːkɪŋ/	n.	金属加工术，金属工
increasingly /ɪnˈkriːsɪŋli/	adv.	日益，越来越多地，不断增加地
precision /prɪˈsɪʒ(ə)n/	n.	精确度，准确（性）
	adj.	精确的，准确的，细致的

indistinguishable /ˌɪndɪˈstɪŋgwɪʃəb(ə)l/		adj.	难区分的，不能分辨的
crucial /ˈkruːʃ(ə)l/		adj.	决定性的，至关重要的，关键性的；极困难的；十字形的，交叉状的
craftsman /ˈkrɑːftsmən/		n.	工匠；手艺人
snag /snæg/		n.	（尤指潜伏的）困难；（未料到的）阻碍，问题，麻烦；尖齿，尖刺；被刮破或钩坏的裂口；根株
		vt.	（被尖锐物）挂住，钩住，刮破；迅速抓到；抢到
stamp /stæmp/		n.	邮票，印花，印，图章；标志
		vt.	贴邮票于；在……盖章；标出；铭记；压印上（图案、标记等）；冲压
jig /dʒɪg/		n.	吉格舞；[机] 夹具，钻模
		vi.	用夹具（钻模）加工
finisher /ˈfɪnɪʃə/		n.	润饰者，完工者
mounter /ˈmaʊntə/		n.	装配工；安装工；镶嵌工
formed /fɔːmd/		adj.	成形的；成形加工的
drill /drɪl/		n.	钻头，钻床；（拍卖时用的）木槌；音槌；击铁
		vt. & vi.	捶打；反复敲打
gravel /ˈgræv(ə)l/		n.	砾石；卵石；砾石层
		vt.	以砾石铺路
fitted /ˈfɪtɪd/		adj.	配备的
progression /prəˈgreʃ(ə)n/		n.	（事件的）连续；一系列；[数] 级数；发展，进程，程序
strip /strɪp/		n.	狭长的一块（材料、土地等）
		vi.	剥皮；（表皮、壳等）剥落；（齿轮的齿等）折断；（螺母、螺栓的螺纹等）磨损
work-in-progress		n.	在制品（完成部分成产程序）
replica /ˈreplɪkə/		n.	复制品，仿制品
assay /əˈseɪ/		n.	化验，试金；分析实验；化验报告；试样；鉴定，测定
		vt.	化验；试验，检验；分析；评价
		vi.	经检验证明内含成分
resurge /rɪˈsɜːdʒ/		v.	复活
electroforming /ɪˈlektrə(ʊ)ˌfɔːmɪŋ/		n.	电铸，电冶，电成型
electrodeposition /ɪˌlektrəʊdepəˈzɪʃn/		n.	电沉积
mandrel /ˈmændr(ə)l/		n.	心轴，芯棒

synthesize /ˈsɪnθəsaɪz/	vt.	综合，使合成；人工合成；
electrolytic /ɪˈlektrə(ʊ)ˈlɪtɪk/	adj.	电解的，由电解产生的
metalize /ˈmet(ə)laɪz/	v.	使金属化，使……硬化
fabricate /ˈfæbrɪkeɪt/	vt.	编造；伪造；装配
forging /ˈfɔːdʒɪŋ/	n.	伪造；锻造；锻件
tolerance /ˈtɒlərəns/	n.	宽容，容忍；忍耐力；耐力；公差
inherent /ɪnˈhɪər(ə)nt/	adj.	固有的；内在的
photographically /ˌfəʊtəˈpræfɪk(ə)li/	adv.	逼真地
patterned /ˈpætənd/	adj.	有图案的，组成图案的
wrought /rɔːt/	adj.	制造的；加工的；经装饰的；锻造的
	vt.	使发生了，造成了（尤指变化）
sterling /ˈstɜːlɪŋ/	n.	英国货币
	adj.	标准纯度的；含标准成分的；品格优秀的
dimensional /dɪˈmenʃ(ə)n(ə)l/	adj.	尺寸的；[物]量纲的；[数]因次的；……维的
scrap /skræp/	n.	小片；少许；废料；吵架；废弃的；丝毫
	vt.	废弃，丢弃；取消；使结题，拆毁
	adj.	零星的；片段的；废弃的，丢弃的
hollow /ˈhɒləʊ/	adj.	空的；空洞的；虚伪的；无真正价值的
	n.	洞；坑；凹地
	vt.	挖洞
lightweight /ˈlaɪtweɪt/	adj.	轻量的，薄型的
	n.	体重低于通常重量的人；比通常重量轻的东西
voluminous /vəˈljuːmɪnəs/	adj.	宽松的，肥大的；容量大的；多卷的；篇幅长的
	adv.	宽松地；容量大地；篇幅长地
	n.	宽松；容量大；篇幅长
unparalleled /ʌnˈpærəleld/	adj.	无比的，无双的，空前的
clasp /klæsp/	n.	扣子，钩环
	vt. & vi.	扣住，扣紧
a strip of		狭长的一条
investment casting		熔模铸造法，失蜡铸造（同 centrifugal casting）
deep drawing		深冲压（金属板坯加工）

Unit 10

Jewelry Commerce

Starting Out

☞ **Match Words with Pictures**

Match the words with the corresponding pictures.

| obsidian | prehnite | moonstone | tourmaline | garnet |
| cat's eye | topaz | rhodochrosite | sugilite | |

1. _____   2. _____   3. _____

4. _____   5. _____   6. _____

7. _____   8. _____   9. _____

☞ **Check Your Knowledge**

Describe the words relating to jewelry commerce.

1. appraise: _____   2. estimate: _____
3. evaluate: _____   4. gem certificate: _____
5. invaluable: _____   6. valueless: _____
7. appreciate: _____   8. brand name: _____
9. trademark: _____   10. price tag: _____

Part I Communicative Activities

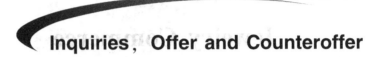

Inquiries, Offer and Counteroffer

Task 1 Conversations

Ⅰ. **Read the following conversation and underline some useful sentences for buying and selling jewelry.**

(D: jewelry dealer; B: buyer)

D: What can I do for you, madam?

B: Well, I hope you can help me. There is something wrong with this diamond ring, would you take a look?

D: Oh, dear. The diamond just comes off its location. Did you drop it or anything?

B: No. I bought it from your shop last week. And I hadn't worn it all the time. But I found the problem this morning.

D: Do you have proof of purchase?

B: The receipt and the guarantee? Here you are.

D: I'm terribly sorry. I apologize for the inconvenience. Now you need to fill this form.

B: Next?

D: We'll have to send it to our factory to fix the diamond again.

B: How long will it be back? I do need it very soon.

D: It will be back in one week.

B: I was hoping you would be able to fix it in good time.

D: Very sorry. But we will treat as soon as possible.

B: Yes, thank you.

D: Once again. We're very sorry about this.

B: Well. I will wait for your word. Goodbye.

D: Goodbye.

Ⅱ. **Look at the following conversation. Decide where the following sentences go and then act out the conversation in pairs.**

a. Half of the price.
b. 15% is off, OK?
c. The price is beyond my budget.
d. Well, is there discount?
e. Why not try on the earrings too?

(D: jewelry dealer; B: buyer)

B: Excuse me. Can I try this emerald necklace on?
D: Yes, sure. It suits you well, I think. 1. _____ They are a whole set of jewelry.
B: The size of the earrings seems too small, isn't it?
D: They are high-grade emeralds from Colombia. The matching is not a problem.
B: Is it of the highest quality?
D: Sure. The craft is also unique.
B: 2. _____
D: How much do you want for this?
B: 3. _____
D: Oh, No. The highest discount is 10% on this set. Our brand is famous, you know.
B: 4. _____ I'll buy it right away if it were cheaper.
D: Let me count. 5. _____
B: Yes. The emerald is attractive. Can you give me the invoice?
D: Sure. Please pay it at the cashier's over there.
B: Thank you.

Task 2　Role-play

Work in pairs and act out the following roles in the conversation about making offers and counteroffers.

Student A: a supplier of jewelry products located in Hong Kong
Student B: a buyer from China

Useful Expressions and Sentences

1. **Making an inquiry**
- There is something wrong with this diamond ring, would you take a look? 这枚钻戒有点问题，你能看看吗？
- Do you have proof of purchase? 你有购货证明吗？
- The receipt and the guarantee? 收据和保修单吗？
- How long will it be back? 要等多久？

2. **Making an offer**
- Well, is there discount? 你有折扣吗？
- How much do you want for this? 你想多少钱要？
- Half of the price. 半价。
- The highest discount is 10% on this set. Our brand is famous, you know. 这款最高折扣是九折。你要知道，我们这是名牌。

3. **Making counteroffer**
- The price is beyond my budget. I'll buy it right away if it were cheaper. 这个价格超过我的预算。你要是能再便宜点，我就立刻买下。
- 15% is off, ok? 八五折，可以吗？

Part II Read and Explore

The Price of Gemstone

The price per carat of different gemstones can vary enormously, literally from $1 a carat to tens of thousands. Many factors can influence the price per carat. Here is a concise summary of the 10 factors that determine gem prices:

Gem Variety

Some gemstone varieties—such as sapphire, ruby, emerald, garnet, tanzanite, spinel and alexandrite command a premium price in the market, due to their superior gemstone characteristics and rarity. Other varieties, such as many types of quartz, are abundant in many locations around the world, so prices are much lower. But while the gem variety sets a general price range for a stone, the characteristics of the specific gem also have a major effect on the price per carat.

Color

In colored gemstones, color is the single most important determinant of value. Ideal colors vary by gem variety of course, but generally the colors that are most highly regarded are intense, vivid and pure. Gems that are too light or too dark are usually less desirable than those of medium tones. Thus a rich cornflower blue in sapphire is more valuable than an inky blue-black or a pale blue.

Clarity

A gemstone that is perfectly clean, with no visible inclusions, will be priced higher. In general, the cleaner the stone, the better it's brilliance. So while it is true that the higher the clarity grade, the higher the value of the gem, inclusions that don't interfere with the brilliance and sparkle of a gem will not affect its value, significantly. Note also that some gems, such as emerald, always have inclusion.

Cut and Polish

Gemstones should be cut with proper proportions to maximize the light that is returned to the eye. But gem cutters or lapidaries often have to make compromises when cutting a particular piece of material. If the gem color is quite light, cutting a deeper stone will provide a richer color. Conversely, a dark tone can be lightened by making a shallower cut. But in every case, the facets should meet cleanly and the surface should be well polished with no scratches.

Size

For some gemstone varieties, such as quartz, the price per carat is fairly constant as the weight of the stone increases. But in the case of many rarer gems, price does not increase in a linear way as the weight increases. Indeed for some gems, such as diamonds, the price per carat can increase exponentially as the carat size increases. According to this formula, a 1 carat stone may cost $1,000 while a 2 carat stone may cost $4,000. Though the formula is rarely so exact, good quality sapphires and rubies in large size do tend to have a much higher per carat price.

Not only are larger stones more expensive, but gems cut in stock sizes—what are known in the trade as calibrated sizes—also tend to be more expensive. This is because more material has to be removed to achieve the calibrated size.

Shape

Some shapes tend to be priced higher than others, in part because of demand and in part because of material issues in cutting the specific shape. In general, round gems tend to command a premium in the market. Rounds are much less common than ovals, since ovals are usually cut to preserve as much weight of the raw material as possible. Cutting a round gem normally requires a greater loss of the rough stone, and for very expensive materials like sapphire, ruby, alexandrite and others, this can have a significant effect on price.

Treatment

Gem treatment such as heating, fracture-filling, radiation and diffusion significantly improve the appearance of many gemstones, and these treatments are now routine for many commercial-grade stones. A treated stone is always less expensive than a similar untreated stone. But most of the stones that are routinely treated—such as ruby and sapphire—are now very rare in untreated form, and the untreated stones fetch a market price out of the reach of most consumers. If you prefer to buy an untreated stone, you do have many choices. A number of popular gems, such as tourmaline, spinel, amethyst and garnet are almost never treated.

Origin

Strictly speaking, a fine natural gem is a fine gem, regardless of its country or region of origin. The reality of the market is that certain gem varieties from locations such as Burma, Kashmir, Sri Lanka and Brazil, command a premium price in the market. It is difficult to say whether this premium is justified, especially with so many fine gems coming from Africa.

Fashion

Some gems, such as blue sapphire, are always in fashion. But some gems become fashionable for short periods, with resulting price increases. Recently we've seen andesine labradorite and diaspore in the spotlight, and strong interest in rutilated quartz. Some very fine gems, such as natural spinel, actually have lower than expected prices because limited supply means that the gems are not promoted heavily in the market.

Supply Chain

The gem trade is a business and everyone in the supply chain—from the mine to the jewelry retailer—is trying to turn a profit. Gemstones can pass through many hands on the way from the mine to the consumer, and the more brokers and distributors that handle the product, the higher the final price will be. So in fact the same gemstone may carry a price that varies by as much as 200%, depending on who you buy it from.

Check Your Understanding

I. Fill in the blanks according to the text.

1. The factors that determine gem prices are _____, _____, _____, _____, _____, _____, _____, _____, _____, _____.
2. In colored gemstones, _____ is the most important determinant of value.
3. Gemstones should be cut with _____ to maximize the light that is returned to the eye.
4. Gem treatments such as _____, _____, _____ and _____ significantly improve the appearance of many gemstones.

II. Answer the following questions.

1. How do we introduce the the variety of gemstone to buyer?
2. Please talk about the color of gemstone.
3. Please talk about the clarity of gemstone.
4. Please talk about the fashion of gemstone.

Subject Focus

1. Write a summary about the price of gemstone and present to the class.
2. Search for information on the price of gemstone and give a presentation about the factors that determine gem prices.
3. Now make a draft of commodity description of a specific item, then post it online and present to the class.

Language Focus

I. Subject-related Terms

Fill in the blanks with the words or expressions being defined.

1. _____ a fact, situation, or experience is unpleasant, depressing or harmful
2. _____ make less visible or unclear
3. _____ allowing light to pass through diffusely
4. _____ lacking solidity or strength and liable to break
5. _____ having a quality that thrusts itself into attention
6. _____ a green transparent form of beryl; highly valued as a gemstone
7. _____ precisely determined, fixed or defined; especially fixed by rule or by a specific and constant cause
8. _____ a highly developed state of perfection; having a flawless or impeccable quality

9. _____ (physics) the process of diffusing; the intermingling of molecules in gases and liquids as a result of random thermal agitation
10. _____ a lamp that produces a strong beam of light to illuminate a restricted area; used to focus attention of a stage performer

II. **Working with Words and Expressions**

In the box below are some of the words and expressions you have learned in this text. Complete the following sentences with them. Change the form of words if necessary.

concise	influence	literally	vary	tend to
have an effect on	set a price for	interfere with	in general	due to

1. We know how the volume and temperature _____ with respect to each other at constant pressure.
2. We've got to get the economy under control or it will _____ eat us up.
3. We became best friends and he _____ me deeply.
4. Be sure to make it clear and _____ and avoid long-windedness.
5. The traffic accident was allegedly _____ negligence.
6. As such, everything is seen as a commodity and if there is no market to _____ then there is no economic value.
7. What you say or do will _____ others.
8. And _____, what you give tend to get back from the world around you.
9. But that is your affair; we do not _____ your problems.
10. We all _____ like those who are similar to us.

III. **Grammar Work**

Observe the following sentences and pay special attention to the use of comparative adjectives.
1. In general, *the cleaner* the stone, *the better* it's brilliance.
2. So while it is true that *the higher* the clarity grade, *the higher* the value of the gem, inclusions that don't interfere with the brilliance and sparkle of a gem will not affect its value, significantly.
3. Gemstones can pass through many hands on the way from the mine to the consumer, and *the more* brokers and distributors that handle the product, *the higher* the final price will be.

Now correct the mistakes in the following sentences.
1. The more he got, the more he wants.
2. The higher the ground is, the thinner air become.
3. The harder you will work, the greater progress you will make.
4. The more English you practised, the better your English is.
5. The busier he is, the happier he felt.

6. The more air there is inside the tyre, the greater pressure there is.
7. The more she will flatter me, the less I like her.
8. The less he worried, the better he has worked.
9. The more annoyed I became, the more happy he was.
10. The more you put your heart into English, the more you've be interested in it.

Ⅳ. Translation

1. The price per carat of different gemstones can vary enormously, literally from $1 a carat to tens of thousands. Many factors can influence the price per carat.
2. Gems that are too light or too dark are usually less desirable than those of medium tones. Thus a rich cornflower blue in sapphire is more valuable than an inky blue-black or a pale blue.
3. So while it is true that the higher the clarity grade, the higher the value of the gem, inclusions that don't interfere with the brilliance and sparkle of a gem will not affect its value, significantly.
4. Strictly speaking, a fine natural gem is a fine gem, regardless of its country or region of origin. The reality of the market is that certain gem varieties from locations such as Burma, Kashmir, Sri Lanka and Brazil, command a premium price in the market.
5. 宝石从开采者手中到最终消费者手中要经过很多次交易，经过的代理和分销商越多，其价格就越高。
6. 圆形宝石较椭圆形不常见，因为椭圆形切工的宝石可以尽量保留原材料的重量。
7. 许多流行的宝石如碧玺、尖晶石、紫晶和石榴石几乎都未经处理。
8. 一些非常好的宝石，比如天然尖晶石，其价格实际上低于预期价格，因为有限的供给使得这些宝石在市场中没有被大规模推广。

Part Ⅲ Extended Learning

 Dictation

Listen to the audio and complete the following passage with the words or expressions you hear.

International Jewelry Fair Ends in Beijing

The 2009 China International Gold, Jewelry, and Gem Fair has 1. _____ a glittering note in Beijing. The fair demonstrated that the Chinese market is attracting considerably more attention than ever before from both 2. _____ suppliers.

Panyu District, in the southern province of Guangdong, is China's biggest 3. _____, accounting for more than 60 percent of the country's export. Jewelry makers there used to concentrate on the international market and rarely participated in 4. _____. This time, however,

23 companies came to Beijing as a group for the fair, after their European sales dropped 5. _____.

Li Zhiwei, Panyu Chamber of Commerce, China International Chamber of Commerce said "Now, 90 percent of our products are exported. Only 10 percent is sold in China. I hope in the next two or three years, the domestic market can 6. _____ 20 percent of our sales."

Jewelry makers say the US market tumbled 60 percent this year, and the Japanese market is still 7. _____. In contrast, the Chinese market has been expanding by 10 percent a year.

Shi Hongyue, Deputy Secretary General of China Jewelry, Jade and Gem Industry Assoc. said, "China's 8. _____ is developing at high speed. Our jewelry industry is developing even faster. Sales this year are expected to hit 9. _____ yuan. And overseas suppliers are enjoying a growing market share."

More than 600 exhibitors from 15 countries and regions attended the mega-event, which presented 10. _____ that included diamond jewelry, gem stones, antique jewelry, pearls and corals, gold and silver jewelry, and precious watches.

Read More

Passage 1

Birthstone

Garnet

Garnet species are found in many colors including red, orange, yellow, green, blue, purple, brown, black, pink and colorless.

The rarest of these is the blue garnet, discovered in the late 1990s in Bekily, Madagascar. It is also found in parts of the United States, Russia and Turkey. It changes color from blue-green in the daylight to purple in incandescent light, as a result of the relatively high amounts of vanadium (V).

Amethyst

Amethyst is a purple variety of quartz (SiO_2) and owes its violet color to irradiation, iron impurities and the presence of trace elements, which result in complex crystal lattice substitutions. The hardness of the mineral is the same as quartz, thus it is suitable for use in jewelry.

Pure quartz, traditionally called rock crystal (sometimes called clear quartz), is colorless and transparent or translucent. Common colored varieties include citrine, rose quartz (pink quartz), prasiolite, rutilated quartz, amethyst, smoky quartz and milky quartz.

Aquamarine

Aquamarine (from Latin: aqua marina, "water of the sea") is a blue or turquoise variety of beryl. It occurs at most localities which yield ordinary beryl.

Common beryl include aquamarine, emerald, goshenite (also known as "colorless beryl"), morganite (also known as "pink beryl" "rose beryl" and "pink emerald"), red beryl (also known as "red emerald" or "scarlet emerald"), golden beryl and heliodor.

Diamond

The most familiar usage of diamonds today is as gemstones used for adornment—a usage which dates back into antiquity. In the twentieth century, four characteristics known informally as the four Cs are now commonly used as the basic descriptors of diamonds: carat, cut, color and clarity.

Emerald

Emerald is a rare and valuable gemstone and, as such, it has provided the incentive for developing synthetic emeralds.

Most emeralds are highly included, so their toughness is classified as generally poor.

Emeralds come from three main emerald mining areas in Colombia.

Pearl

Freshwater and saltwater pearls may sometimes look quite similar, but they come from different sources.

Freshwater pearls form in various species of freshwater mussels, family Unionidae, which live in lakes, rivers, ponds and other bodies of fresh water. However, most freshwater cultured pearls sold today come from China.

Saltwater pearls grow within pearl oysters, family Pteriidae, which live in oceans. Saltwater pearl oysters are usually cultivated in protected lagoons or volcanic atolls.

Ruby

A ruby is a pink to blood-red colored gemstone, a variety of the mineral corundum (aluminium oxide Al_2O_3). The red color is caused mainly by the presence of the element chromium. Its name comes from ruber, Latin for red. Other varieties of gem-quality corundum are called sapphires. The ruby is considered one of the four precious stones, together with the sapphire, the emerald, and the diamond.

Prices of rubies are primarily determined by color. The brightest and most valuable "red" called pigeon blood-red, commands a huge premium over other rubies of similar quality.

Ruby under a normal light (top) and under a green laser light (bottom). Red light is emitted.

Peridot

Peridot is one of the few gemstones that occur in only one color, an olive green. The intensity and tint of the green, however, depends on how much iron is contained in the crystal structure, so the color of individual peridot gems can vary from yellow to olive to brownish-green. The most valued color is a dark olive-green.

Sapphire

Sapphire (Greek: "blue stone") is a gemstone variety of the mineral corundum, an aluminium oxide (Al_2O_3). Trace amounts of other elements such as iron (Fe), titanium (Ti), or chromium (Cr) can give corundum blue, yellow, pink, purple, orange, or greenish color. Chromium (Cr) impurities in corundum yield a red tint, and the resultant gemstone is called a ruby.

A star sapphire is a type of sapphire that exhibits a star-like phenomenon known as asterism; red stones are known as "star rubies".

Opal

Opal is the national gemstone of Australia, which produces 97% of the world's supply.

Opal's internal structure makes it diffract light; depending on the conditions in which it formed it can take on many colors. Opal ranges from clear through white, gray, red, orange, yellow, green, blue, magenta, rose, pink, slate, olive, brown, and black. Of these hues, the reds against black are the most rare, whereas white and greens are the most common.

The Flame Queen Opal is perhaps the most famous of all opals.

Topaz

Pure topaz is colorless and transparent but is usually tinted by impurities. Typical topaz is wine, yellow, pale gray, reddish-orange, or blue brown. It can also be made white, pale green, blue, gold, pink (rare), reddish-yellow or opaque to transparent/translucent.

Imperial topaz is yellow, pink (rare, if natural) or pink-orange.

Blue topaz is the US state Texas' gemstone. Naturally occurring blue topaz is quite rare.

Mystic topaz is colorless topaz which has been artificially coated giving it the desired rainbow effect.

Turquoise

The substance has been known by many names, but the word turquoise, which dates to the 16th century, is derived from an old French word for "Turkish".

It is rare and valuable in finer grades and has been prized as a gem and ornamental stone for thousands of years owing to its unique hue.

Passage 2

About GIA and Its Grading Fancy-Color Diamonds

About GIA

GIA is the largest, most respected nonprofit source of gemological knowledge in the world.

GIA exists to connect people to the understanding of gems. As a long-standing scientific authority, GIA is not only a unique source for diamond knowledge, its grading reports inspire confidence wherever they appear.

GIA's commitment to protecting diamond buyers inspired the Institute to create the 4Cs and the International Diamond Grading System^{TM}. These methods are the universal benchmarks by which all diamonds are judged. As the birthplace of these standards, and with its investment in continued gemological research, GIA's authority is unequaled.

The world's most respected retailers, museums, auction houses and private collectors rely on the expertise of GIA graders to assess, grade and verify their gems. They recognize the importance of complete, unbiased, scientific information in gem assessment, and absolutely trust GIA to provide it.

Fancy Color Diamonds and Gemstones

While the vast majority of diamonds fall in the D-to-Z color range, nature occasionally produces diamonds with a naturally occurring blue, brown, pink, deep yellow or even green hue. The geological conditions required to yield these colors are rare, making diamonds with distinct and naturally occurring shades scarce and highly prized.

Unlike colorless and near-colorless diamonds, fancy-color diamonds are evaluated less for brilliance or fire and more for color intensity. Shades that are deep and distinct are rated higher than weak or pale shades.

GIA describes color in terms of hue, tone and saturation. Hue refers to the diamond's characteristic color, tone refers to the color's relative lightness or darkness and saturation refers the color's depth or strength. Using highly controlled viewing conditions and color comparators, a fancy color grader selects one of 27 hues, then describes tone and saturation with terms such as "Fancy Light" "Fancy Intense" and "Fancy Vivid". The color system GIA developed is used worldwide.

Colored Gemstone Identification

While renowned for its diamond grading expertise, GIA also receives a vast array of colored gemstones for identification. Over the decades, the Institute has created a database of information on more than 100,000 individual colored gemstones. Using this database and sophisticated analytical tools, GIA graders and researchers can pinpoint a gem's identity and, depending on the gemstone, its geographic origin. They also identify synthetics, simulants, and stones that have undergone treatment. A particularly important activity is determining whether a gemstone's color is natural or the result of a treatment process.

GIA's processes for evaluating colored stones involve multiple graders and the same item identification and tracking procedures used in diamond grading.

GIA offers two types of grading report for colored diamonds. The GIA Colored Diamond Grading Report contains the same comprehensive diamond 4Cs information as the GIA Diamond Grading Report, while the GIA Colored Diamond Identification and Origin Report (also known as the color-only report) is limited to color grade and the origin of the color (natural or treated).

Pearl Classification and Grading

For more than 100 years, discoveries in pearl culturing have revolutionized the market and essentially replaced natural pearls in jewelry.

A natural pearl occurs when an irritant, such as a parasite, works its way into a particular species of oyster, mussel or clam. In defense, the mollusk secretes fluid, called nacre, to coat the irritation. Over time, layers of nacre form natural pearl. In cultured pearls, the irritant is a surgically implanted bead or bit of shell.

The ability to consistently generate what was once a rare phenomenon has created a much wider audience for the appreciation and purchase of pearls. But it has also led to confusion about levels of quality and how to determine them. Cultured pearls are produced in a wide variety of colors, shapes, and sizes and grading has become proportionately complex.

In response, GIA has created a standard for describing pearl quality just as it did with diamonds. GIA's system, launched in 1998, is based on 7 Pearl Value Factors™: size, shape, color, luster, surface quality, nacre quality, and matching.

Group Project

1. Choose one of the Successful Jewelry E-Commerce and write a passage about it, giving as many details as possible.
2. Compare the differences between the Chinese people and Westerners in viewing and treating Jewelry E-Commerce, and give a presentation on it.
3. Study the current development in the Jewelry E-Commerce and write a report about it.

Words and Expressions

estimate /ˈestɪmeɪt/	v.	估计，估价，评估
	n.	评价，估计，估价，评估
evaluate /ɪˈvæljʊeɪt/	vt.	评价，估计，求……的值
	v.	评价
certificate /səˈrtɪfɪkət/	n.	证书，证明书
	vt.	发给证明书，以证书形式授权给……
invaluable /ɪnˈvæljʊb(ə)l/	adj.	无价的，价值无法衡量的
valueless /ˈvæljʊləs/	adj.	不足道的
appreciate /əˈpriːʃɪeɪt/	vt.	赏识，鉴赏，感激
	vi.	增值，涨价
trademark /ˈtreɪdmɑːk/	n.	商标
enormously /ɪˈnɔːməsli/	adv.	非常，巨大地
literally /ˈlɪtərəli/	adv.	照字面意义，逐字地
concise /kənˈsaɪs/	adj.	简明的，简练的
command /kəˈmɑːnd/	n.	命令，掌握，司令部
	v.	命令，指挥，克制，支配
premium /ˈpriːmɪəm/	n.	额外费用，奖金，奖赏，保险费
quartz /kwɔːts/	n.	石英
determinant /dɪˈtɜːmɪnənt/	adj.	决定性的
intense /ɪnˈtens/	adj.	强烈的，剧烈的，热切的，热情的
cornflower /ˈkɔːnflaʊə/	n.	[植] 矢车菊
inky /ˈɪŋki/	adj.	漆黑的，墨水的，给墨水弄污的
clarity /ˈklærɪti/	n.	清楚，透明
polish /ˈpɒlɪʃ/	n.	磨光，光泽，上光剂，优雅，精良
	vt.	擦亮，发亮，磨光，推敲，发亮，变光滑
proportion /prəˈpɔːʃ(ə)n/	n.	比例，均衡，面积，部分
	vt.	使成比例，使均衡，分摊

lapidary /ˈlæpɪdəri/	n.	宝石鉴定家
compromise /ˈkɒmprəmaɪz/	n.	妥协，折中
	v.	妥协，折中，危及……的安全
conversely /ˈkɒnvɜːsli/	adv.	相反地
scratch /skrætʃ/	n.	乱写，刮擦声，抓痕，擦伤
	vt.	乱涂，勾抹掉，擦，刮，搔，挖出
	vi.	发刮擦声，搔，抓
	adj.	打草稿用的，凑合的
exponentially /ˌekspəˈnenʃ(ə)li/	adv.	以指数方式
stock size /stɒk saɪz/		（鞋、帽、成衣等的）现货号码（尺寸）
calibrate /ˈkælɪbreɪt/	vt.	校准
loss /lɒs/	n.	损失，遗失，失败，输，浪费，错过
diffusion /dɪˈfjuːʒ(ə)n/	n.	扩散，传播，散射
gem certificate		宝石鉴定证书
brand name		商标，品牌
price tag		价格标签
visible inclusion		可见内含物
commercial-grade stone		商业级宝石

Key to the Exercises and Translation

Unit 1 Introduction to Gemology

Starting Out

◆ Match Words with Pictures

1. gold 2. crystal 3. pearl 4. jadeite 5. ruby 6. diamond

◆ Check Your Knowledge

1. 镀金   2. 白金   3. 铜   4. 锆石   5. 合金
6. 银   7. 猫眼石   8. 亚克力   9. 玛瑙   10. 碧玺
11. 发夹   12. 吊坠   13. 胸针   14. 手镯   15. 脚链
16. 手链   17. 项链   18. 套饰   19. 皮带扣   20. 丝巾扣

Part Ⅰ Communicative Activities

1. e 2. c 3. b 4. f 5. d 6. a

Part Ⅱ Read and Explore

◆ Check Your Understanding

Ⅰ.

look/style	bracelet	earrings	necklace
casual look	silver or gold bangles/a wide metal or acrylic slip-on bracelet	hoop earrings/diamond or other gem studs	a precious gem, cystal or rhinestone charm
edgy look	a metal cuff bracelet		
formal or upscale look	an elegant and simple diamond tennis bracelet/a mesh bracelet made of a precious metal	drop earrings	a precious gem, cystal or rhinestone charm
a V-neck or scoop-neck top			a drop necklace with a charm on the end
strapless dresses/spaghetti strap styles			choker necklaces

Ⅱ. 1. The most important principle in wearing fashion jewelry is less is more.
   2. The author recommends us to wear as one piece of jewelry as possible in one look.
   3. The case when bracelet, necklace, and earrings are acceptable in one whole outfit is that you have a simple drop diamond charm necklace, diamond stud earrings and a tennis bracelet.
   4. They differ in shapes and occasions when they are suitable to be worn.
   5. Wearing bracelets on both wrists is not preferred.

◆ Language Focus

Ⅰ. 1. neckline 2. gem 3. charm 4. accessory 5. outfit

English for Jewelry

  6. gown   7. rhinestone   8. acrylic   9. top   10. crystal

Ⅱ. 1. extravagant   2. goes; with   3. head-turning   4. Casual   5. overdoing
  6. feel free to   7. kept in mind   8. stick with   9. neglected   10. overwhelmed

Ⅲ. 1. 去掉 it   2. which 改为 where   3. Which 改为 As   4. practice 改为 practices
  5. who's 改为 whose   6. whom 改为 who   7. who his 改为 whose   8. which 改为 when
  9. which 改为 wh… 10. 去掉 there

Ⅳ. 1. 穿着休闲西裤、牛仔裤或女装衬衫，甚至是你的黑色的小礼服时，千万别忘了做最后的装饰。
  2. 如果着装时尚前卫，例如不对称的短裙，可以搭配金属开口镯（形状像字母 C）。
  3. 如果穿着正装或者高级时装，则可以佩戴优雅简约的块状钻石手链。
  4. 大多数款式的吊坠耳环都能搭配漂亮的晚礼服（这种搭配通常出现在红毯活动上）、考究的西裤或牛仔裤，甚至是穿着回头率很高的上装。
  5. If you go for a casual look, try jeans, white short-sleeve blouse, matched with a gold bracelet.
  6. Wearing too many and too complicated jewelry accessories will overwhelm the whole look.
  7. There are mainly three types of earrings: hoops, drops and studs.
  8. The principle of wearing jewelry to be kept in mind is less is more.

◆ Text（Translation）

女性珠宝首饰选择贴士

  穿着休闲西裤、牛仔裤、女装衬衫，甚至是你的黑色小礼服时，千万别忘了做最后的装饰。珠宝首饰是时装不可或缺的重要组成部分。即便仅在拇指上佩戴一枚金戒指，整体气质也会大不一样。穿衣打扮的时候，我们需要牢记一些珠宝首饰选择小贴士。

手镯

  穿着女装衬衫和短袖或七分袖衬衫时，应该佩戴手镯。如果身着便装，例如牛仔裤，则可以佩戴银手镯、金手镯、较宽的金属手镯或方便佩戴的亚克力手镯。如果着装时尚前卫，例如不对称的短裙，可以搭配金属开口镯（形状像字母 C）。如果穿着正装或者高级时装，则可以佩戴优雅简约的块状钻石手链或贵金属材质的网格手链。尽量避免在两个手腕上同时佩戴手镯，否则有可能会影响整体观感。

项链

  项链的样式主要取决于女装衬衫或裙子的领口款式。如果穿着 V 领或圆领上衣，则可以佩戴带有水滴吊坠的项链。正装和休闲装可以搭配价值连城的宝石、水晶或水钻。露肩和低胸礼服则最好搭配贴颈项链。

耳环

  细节决定一切。即便是最小的饰品，哪怕是一个钻石耳钉，也可以与你的着装形成完美的搭配效果。如今主要流行三类耳环，一种是环状耳环，另一种是吊坠耳环，还有一种就是耳钉。

  环状耳环适宜搭配休闲装。如果环状耳环较小（宽度大约与手指宽度相同），则可以搭配正装，尤其是镶有钻石和人造钻石的环状耳环。吊坠耳环一般属于奢侈品或艺术品，注定会为佩戴者赢得赞美。出席特殊场合而你又是此类活动的核心人物时，可以佩戴吊坠耳环。大多数款式的吊坠耳环都能搭配漂亮的晚礼服（这种搭配通常出现在红毯活动上）、考究的西裤或牛仔裤，甚至是穿着回头率很高的上装。镶嵌钻石或宝石的耳钉则适宜搭配休闲装和日常服饰。

物以稀为贵

  不要佩戴过多的珠宝首饰。根据自己的着装风格选择首饰，有时仅仅需要搭配一件有趣的小饰品，例如多彩的水晶手镯或者一对吊灯形耳环，就可以使自己大放异彩。如果你拥有一条简约的水晶吊坠项链，水晶耳钉和块状手镯，则是你唯一可以同时佩戴，以做装饰的搭配，其他情况则不要佩戴过多首饰。

Key to the Exercises and Translation

Part Ⅲ Extended Learning

◆ Dictation

1. trends   2. stunning   3. crystals   4. neckline   5. favorites
6. shoulder pads   7. over-the-top gowns 8. backless   9. stylist   10. glamorous

◆ Read More（Translation）

Passage 1

时尚感十足的六大明星

人们通常以社会名流的装扮作为时尚风向标，因为社会名流的装扮饰品较为前卫，往往在产品上市销售之前就已经穿戴在身。他们的着装打扮会影响粉丝的购买欲望，影响零售商的销售状况。

凯特·米德尔顿给世人留下了深刻的印象，人们无法忘记她优雅的女性气质。她着装打扮端庄大方，风格与其实际年龄相适宜。拍摄订婚照时，她身着手工制作的海军蓝裹身裙，而这一装扮使得裹身裙颇为畅销。她结婚时所穿的蕾丝婚纱就是当下蕾丝裙的前身。此外，她热衷于收集定制外套，这使得带有束腰的长款外套极为畅销。

艾玛·沃特森还是十几岁的孩子时，就被看作时尚偶像。她喜欢香奈儿巧妙的设计风格。她的服饰通常采用空灵而又飘逸的面料。她的便装衣柜里有时装定制围巾和短夹克。

戴安·克鲁格着装风格独特。她能够穿出服饰的时尚气息，而极少有人能穿出如此风格。她棱角分明，皮肤白皙，犹如陶瓷一般。她是戛纳电影节上最耀眼的明星之一，经常参加各种顶级时尚活动。

佐伊·丹斯切尔在那些热爱时装的女性眼里就是完美的时尚典范。她穿衣风格简约，显示出自己可爱、古灵精怪的一面，风格易于模仿。A字裙可以彰显身体的每个细节，而头巾又可以增加她的少女气息。佐伊喜欢穿平底鞋，很少穿高跟鞋。整体来说，她的穿衣风格平易近人。

安德鲁·加菲尔德似乎不太能够成为时尚明星，但他对于时尚的品位透露着成熟的气息。他喜欢穿着修身休闲裤和颜色鲜明的运动上衣以及专业的运动鞋。虽已成年，但其穿着却颇显年轻。

尼娜·多布雷夫长相秀气，曾经出演《吸血鬼日记》。她在时尚界占有一席之地，她偏爱大胆的印花图案，搭配明亮的霓虹色来弥补自己肤色上的不足。她的服饰剪裁较为简单，但却能够凸显她苗条的身材。她颇具时尚眼光。她头发呈波浪状，身着休闲短裙，头发向后盘成发髻，以便凸显晚礼服的线条。尼娜目前备受瞩目。希望她在职业生涯中能够继续大放异彩。

Passage 2

节日礼品建议：首字母饰品

首字母首饰可以作为您的个性化节日礼物！其作为节日礼品可谓是方便又有心思。1928珠宝首饰店出售各式各样的饰品，包括14K首字母黄金耳钉，字母钥匙链和首字母钻石吊坠项链。

华丽的祖传盒式吊坠项链

如果您想准备一份特殊的节日礼物，这款古色古香的祖传盒式吊坠项链是您的首选。作为收藏品的32英寸长古式黄金色调绳链的末端是一个盒式吊坠。吊坠为3.5英寸长，1.5英寸宽，中间镶嵌有一颗红宝石，而周围的花卉图案中镶嵌有施华洛世奇水晶。这款项链在美国制作完成，配有灰绿色的饰品袋。

午夜金光黑饰品套装

是否希望为即将到来的秋天寻找一款前卫的饰品？当你看到这套耀眼的设计师签名版午夜黄金饰品时，我们知道你也希望得到这样一套饰品。整套饰品采用喷金焊接工艺，包括独特的耳环、项链和手镯，尽显美丽大方。这些饰品让我们想起浪漫星空下的夜晚或在空中鸟瞰天色渐暗时摩天大楼亮灯的情景。

应季流行色：灰色如黑色一样经典

又到了灰色流行的季节，这是一种像黑色一般经典百搭的颜色！你可选择羽毛般的淡灰色，也可以选择木炭般的深灰色，灰色让人着迷之处正是其作为单色调配色时的灵活性。

珠宝首饰：哪种金属色适宜搭配灰色？

灰色适合搭配银色珠宝首饰。而在黄金首饰中加入灰色/银色或混合金属元素，则可以透露出独特而又奢华的气息。灰色与金色搭配的戴安娜珠宝首饰系列正是如此。佩戴之后方知实际效果……

Unit 2 Characteristics and Classification of Gems

Starting Out

◆ **Match Words with Pictures**

1. amber 2. coral 3. diamond 4. emerald 5. pearl 6. quartz 7. ruby 8. opal 9. jadeite
10. turquoise 11. chrysoberyl 12. peridot

◆ **Check Your Knowledge**

1. 宝石   2. 半宝石   3. 化学成分   4. 晶体结构   5. 晶系   6. 非晶质体
7. 光学效应 8. 有机宝石 9. 光学性质   10. 颜色分布   11. 透明度   12. 合成宝石

Part Ⅰ Communicative Activities

1. c 2. a 3. e 4. b 5. d

Part Ⅱ Read and Explore

◆ **Check Your Understanding**

Ⅰ. 1. group; type; variety

2. chemical composition; crystal structure; physical qualities

3. single structure; amorphous structure

4. optical qualities

5. jet; coral; ivory; pearl; amber

Ⅱ. 1. There are 16 gemstone groups and more than 130 gemstone types.

2. Optical qualities include color, optical phenomena, color distribution and transparency.

3. There is iridescence, chatoyancy, adularescence, aventurescence, asterism and color change.

4. Gemstones can be transparent, translucent or opaque.

5. When compared to their naturally occurring counterparts, synthetic gemstones have identical chemical composition and crystal structure, including specific gravity and other various optical qualities. But simulated gems are simply imitations possessing only optical similarities.

◆ **Language Focus**

Ⅰ. 1. gemstone   2. crystal structure   3. chemical composition   4. optical phenomena
   5. transparency   6. iridescence   7. chatoyancy   8. organic gems
   9. synthetic gems 10. specific gravity

Ⅱ. 1. available   2. in actuality   3. possessed   4. distinguish   5. transmitted
   6. classify   7. confused with   8. perceive   9. complex   10. define

Ⅲ. 1. C 2. A 3. B 4. A 5. C 6. D 7. C 8. A 9. B 10. B

Ⅳ. 1. 宝石分类的第一步是根据晶体结构和相关的化学物质来区分各种宝石族。

2. 不过这些宝石种都有多种变体，因此将其归为族是合理的。

3. 目前市场上有130多组宝石种，新发现的品种仍不断增加。

4. 在宝石业中，透明度指的是宝石的透光能力，这是宝石常用的分类标准。

5. A gem, or gemstone, is a type of material that is capable of being cut and polished for use in jewelry or other ornamental applications.

6. Diamonds, emeralds, rubies and sapphires are all considered to be precious, and thereby the most valu-

able and most desirable.

7. Ruby is the red variety of corundum, while any other color of corundum is considered sapphire.
8. Gems can be formed from minerals, organic materials and other inorganic materials.

◆ Text（Translation）

宝石分类方法

你是否想过宝石是怎么分类的？根据矿物学理论，宝石分类的第一步是根据晶体结构和相关的化学物质来区分各种宝石族。宝石族可再分成不同的种，种再细分为类。这个过程看似简单，但是，实际上，大多数普通消费者并不了解宝石的族、种、类其实是不同层次的宝石分类。该分类方法由毕业于美国宝石研究院的高纳略·小赫尔伯特与罗伯特·凯莫林确立，显然，他们的分类体系至今仍为宝石学家所使用。

宝石族

族指的是具有相似的化学成分、晶体结构或物理性质的两种或两种以上的宝石。宝石共有16组矿物族，包括绿宝石、金绿宝石、刚玉、钻石、长石、石榴石、玉石、天青石、欧泊、橄榄石、石英、尖晶石、托帕石、碧玺、绿松石和锆石。有意思的是，有几组宝石族实际上是独立的宝石种，如碧玺、锆石、托帕石和尖晶石。不过这些宝石种都有多种变体，因此将其归为族是合理的。

宝石种

宝石种有着变化的化学成分和晶体结构。目前市场上有130多组宝石种，新发现的品种仍不断增加。宝石化学成分相差甚远，有多种成分构成的复杂组合，也有简单、单一的化学元素，比如钻石的成分只有碳。尽管钻石的构成简单，其晶体结构却相当复杂。晶体结构有简单的单晶结构，也有极其复杂的微小晶体簇。有些宝石种甚至可能属于非结晶结构，比如欧泊，这种称为非晶质结构。非有机宝石按照化学成分和晶体结构相似性进行分类，而有机宝石的分类依据只有化学成分。属于有机宝石的有珍珠、珊瑚、琥珀和象牙。

宝石类

宝石种进一步细分为类。最典型的例子是刚玉，刚玉是一个宝石族，蓝色刚玉俗称蓝宝石，红色刚玉即红宝石，它们分别自成一组宝石种。有星光的蓝宝石是蓝宝石中的一类，称为星彩蓝宝石。宝石类别按照光学性质区分，包括颜色、光学效应、颜色分布和透明度。

颜色：颜色是由一定频率的光的传播和吸收所产生的。对颜色的描述需要结合不同的色彩、色调和强度。通过颜色区分类别的宝石如黄水晶，属石英的金黄色类别，又如紫水晶，为石英的紫色类别。

光学效应：最常见的光学效应是晕彩效应。晕彩效应的特征包括珍珠光泽、变彩效应和拉长石晕彩。通过光学效应区分类别的宝石如火玛瑙，它只是具晕彩效应的褐色玛瑙，由于光折射过氧化铁的薄层，呈现红、金、绿、紫罗兰等颜色。有些宝石根据猫眼效应区分类别，如猫眼碧玺或石英猫眼石。其他的光学效应包括光彩效应、砂金效应、星光效应和变色效应。

颜色分布：颜色分布指纯色和纹路在宝石上的分布。有些宝石有着独特的单色和纹路分布，需要单独使用一个类别名称来恰当区分。例如缟玛瑙实际上是有平行排列纹路而非曲线纹路的玛瑙。

透明度：在宝石业中，透明度指的是宝石的透光能力，这是宝石常用的分类标准。宝石分为透明、半透明和不透明。无色石英是透明类石英，而玛瑙和月光石是半透明。碧玉和虎眼石都被认为是不透明的，因为这类宝石不透光。

有机宝石和合成宝石

有机宝石是生物作用自然形成的宝石。有机宝石种类不多，包括煤玉、珊瑚、象牙、珍珠和琥珀。

合成宝石是在实验室环境下生成的。合成宝石具有天然的成分，但不是自然生成的。跟天然宝石相比，合成宝石有着相同的化学成分和晶体结构，包括比重和其他各种光学性质。然而，必须指出的是，并非所有实验室制造的宝石都是合成宝石。这是因为有些实验室生成的宝石使用了非天然的成分，如实验室生成的欧泊，其组成成分是 70%～80% 的硅和 20%～30% 的黏结剂。

人们经常混淆合成宝石和仿制宝石，但是仿制宝石只是拥有相似的光学性质的仿制品。立方氧化锆作为一种仿制钻石是最典型的仿制宝石例子，立方氧化锆看似与天然钻石无异，其化学成分和晶体结构却有着天壤之别。

Part Ⅲ　Extended Learning

◆ **Dictation**

1. anything else　　2. come in any width　　3. version　　4. pattern
5. sapphires　　6. an engagement ring　　7. emerald　　8. square bands
9. newly made　　10. ordinary

◆ **Read More**（Translation）

Passage 1

诞生宝石和星座宝石

宝石爱好者对诞生宝石有着特别的兴趣，宝石与月份的联系可以追溯到数百年前。著名宝石学家 G. F. 昆兹在他的书《宝石奇志》中指出："人们选择嵌有诞生宝石的戒指或饰品，完全是因为乐于相信该宝石与自己的性格密切相连，其他宝石纵使美艳夺目、价值连城也无法比拟。宝石拥有某种难以形容却又真实存在的意义，这样的观念由来已久，对于生性浪漫、富于想象的人来说依然魔力不减。"

最先用于对应星座的是贵重宝石。现今使用的诞生宝石是美国珠宝零售商协会在 1912 年确定的。

诞生宝石和星座石信息

1 月的诞生宝石——石榴石

水瓶座的星座石（1 月 21 日—2 月 18 日）：石榴石

石榴石具有保护佩戴者避免噩梦和蛇咬的力量，人们认为它能在黑暗中指明方向。传统上该石与血液密切相关。经常被作为结婚 2 周年纪念日的礼物。

2 月的诞生宝石——紫水晶

双鱼座的星座石（2 月 19 日—3 月 20 日）：紫水晶

人们认为紫水晶能带来平静、安宁和节制。另外它还能控制性欲和酗酒，这一点可能是优点，也可能是缺点。人们还相信紫水晶可以改善肌肤和发质，预防脱发。据说它能保护佩戴者免遭欺罔。其紫色象征着皇族，是结婚 6 周年纪念日的理想礼物。

3 月的诞生宝石——海蓝宝石

白羊座的星座石（3 月 21 日—4 月 20 日）：血精石

海蓝宝石象征着健康、爱情、年轻和希望。水手笃信它的保护力量。海蓝宝石跟绿宝石同属绿宝石族，经常作为结婚 19 周年纪念日礼物。血精石是对应的星座石。

4 月的诞生宝石——钻石

金牛座的星座石（4 月 21 日—5 月 21 日）：蓝宝石

钻石因其坚硬无比，向来是爱情和永恒的象征。钻石一词来源于希腊文的"金刚石"，意指不可征服。传统上，钻石被用作结婚 10 周年和 60 周年纪念日的礼物。蓝宝石是金牛座的星座石。

5 月的诞生宝石——祖母绿

双子座的星座石（5 月 22 日—6 月 21 日）：玛瑙

据说祖母绿能治愈健康欠佳者，治疗各种疾病，保佩戴者安康。人们还相信佩戴祖母绿能预见未来。对应的星座石是玛瑙。

6 月的诞生宝石——珍珠、月光石

巨蟹座的星座石（6 月 22 日—7 月 22 日）：祖母绿

浪漫的珍珠象征纯洁和端庄。据说珍珠可以使婚姻稳固。对应的星座石是祖母绿。

7月的诞生宝石——红宝石

狮子座的星座石（7月23日—8月23日）：缟玛瑙

人们相信红宝石能使佩戴者生活和谐，佩戴它可保安宁。对应的星座石是缟玛瑙。

8月的诞生宝石——橄榄石

处女座的星座石（8月24日—9月22日）：红玛瑙

橄榄石被认为能避邪，尤其是避开黑暗的妖力。它还能增强药物的疗效。对应的星座石是红玛瑙。

9月的诞生宝石——蓝宝石

天秤座的星座石（9月23日—10月23日）：贵橄榄石

佩戴蓝宝石可享忠贞和纯洁。人们普遍认为蓝宝石越鲜艳越好，可赋予幸运的佩戴者洞察未来的能力。对应的星座石是贵橄榄石。

10月的诞生宝石——欧泊、碧玺

天蝎座的星座石（10月24日—11月22日）：绿宝石

据说欧泊是希望之石，象征纯洁无邪。欧泊可促进康复，增进友谊和心理健康。绿宝石是该星座的星座石。

11月的诞生宝石——黄玉、黄水晶

射手座的星座石（11月23日—12月21日）：黄水晶

据说冷色的黄玉有治疗疾病、保持清醒的效果，象征生命和身心的力量。以前人们曾相信当危险临近时佩戴者可隐身。对应的星座石是黄水晶。

12月的诞生宝石——绿松石、蓝托帕石、坦桑石、蓝锆石

摩羯座的星座石（12月22日—1月20日）：红宝石

人们相信绿松石可以带来福气、好运和幸福，令人心想事成。红宝石是该星座的星座石。

Passage 2

在售商品：浅绿橄榄石

2.5克拉浅绿橄榄石　9毫米×8毫米

品种：橄榄石	商品号：334417
含量：1件	重量：2.5克拉
尺寸：8.9毫米×7.87毫米×4.88毫米	形状：枕形棋盘式
净度：VS—SI	处理：无
产地：中国	价格：90.00美元

保证100%天然宝石

以上图片为出售实物照片。

商品描述

该天然浅绿橄榄石产自中国，重约2.5克拉。尺寸为8.9毫米×7.87毫米×4.88毫米（长×宽×高）。形状（切割方式）是枕形棋盘式。该2.5克拉橄榄石现接受订购，可运至全球各地。可提供宝石证书，此为自选服务项目。

运费

- FedEx或DHL快递仅20美元/运送时间3~4天/全保。
- 挂号件6.99美元/运送时间7~14天/保额200.00美元。

付款方式

- 信用卡
- 贝宝

- 西联汇款
- 银行转账

保证/证书
- 所有的宝石都是天然的。
- 所有照片和视频为实物拍摄（非图库照片）。
- 我们提供 AIGS 证书，费用 30 美元（需多花 5~10 天）。
- 我们还提供 BGL 证书，费用仅 15 美元（需 3~4 天）。

尺寸和重量
- 宝石测量单位一般为毫米（mm）。
- 尺寸表示方式为长×宽×高，圆形宝石尺寸表示方式为直径×高。
- **宝石的挑选看尺寸而非重量！** 不同宝石类别在比重方面有所不同，因此克拉重量不能准确体现其尺寸。
- 注：1 克拉 = 0.2 克

追加数量
- 每款商品都是独一无二的，每款只能订购一件。
- 但是我们经常有类似或相配的商品。如果有的话会在该款宝石的同一页面展示出来。

净度详解
- IF = 内部无瑕级——内部无瑕疵；无内含物。
- VVS = 极微瑕——内含物极其微小；用 10 倍放大镜观察极难看见内含物。
- VS = 微瑕——内含物非常微小；用 10 倍放大镜观察方能看见内含物。
- SI = 小瑕——肉眼可见微小内含物。
- I1 = 一级瑕——肉眼可见内含物。
- Transparent = 宝石可清晰地透光/可能含有金红石或其他内含物。
- Translucent = 可透光，但不透明。
- Opaque = 不透光。

退货规定
- 所有宝石自发货日起享 30 天观察期，在此期内可全额退货。
- 退货时请发邮件给我们申请退货授权码。

橄榄石简介

　　橄榄石是镁-铁橄榄石矿物系列的成员，属橄榄石族，是一种自色宝石。自色宝石指宝石的颜色由矿物本身的基本化学成分而非微小杂质所致，因此，橄榄石的颜色只有绿色。事实上，橄榄石是极少数单一颜色宝石之一。历史上埃及是主要的产地，而今天已经被亚利桑那和巴基斯坦取代。巴基斯坦橄榄石品质出众，20 世纪 90 年代中期在巴基斯坦新发现的矿源拓宽了橄榄石的市场。中国的橄榄石供应量也在增长，我们最近还从越南采购到上乘的橄榄石。

Unit 3　Famous Jewelry and Their History

Starting Out

◆ **Match Words with Pictures**

1. bracelet　2. pendant　3. necklace

◆ **Check Your Knowledge**

The Heart of the Ocean Pendant—*Titanic*

Ruby and Diamond Necklace—*Pretty Woman*

Key to the Exercises and Translation

The One Ring—*The Lord of the Rings*
Audrey Hepburn's Pearls—*Breakfast at Tiffany's*
The Baseball Diamond—*The Great Muppet Caper*
Diamond Necklace—*To Catch a Thief*

Part Ⅰ Communicative Activities

1. b   2. c   3. a   4. d

Part Ⅱ Read and Explore

◆ **Check Your Understanding**

Ⅰ. 1. C   2. E   3. F   4. B   5. A   6. D   7. G

Ⅱ. 1. No, it does not with Taylor.
   2. Harry Winston and his cleaver, Pastor Colon Jr. studied it for six months. Markings were made, erased and redrawn to show where the stone could be cleaved.
   3. No, Burton was determined to acquire the diamond and spoke to the agent.
   4. Taylor wore the diamond to attend Princess Grace's 40th birthday party in Monaco on November 12th.
   5. It now weighs 68.09 carats.

◆ **Language Focus**

Ⅰ. 1. carat   2. auction   3. bidder   4. agent   5. proceeds

Ⅱ. 1. outrageous   2. displayed   3. acquired   4. enhance   5. applicable
   6. erupted   7. comment   8. exclaiming   9. simultaneous   10. previous

Ⅲ. 1. light-heart 改为 light-hearted   2. good-looked 改为 good-looking
   3. long-stand 改为 long-standing   4. cutting-price 改为 cut-price
   5. Free-duty 改为 Duty-free   6. second-classes 改为 second-class
   7. lastly-minute 改为 last-minute   8. grow-up 改为 grown-up
   9. five-pages 改为 five-page   10. down-and-earth 改为 down-to-earth

Ⅳ. 1. 现在对一些女性来说可能就是这样——通常佩戴着大得夸张的珠宝首饰——在一定程度上显露了她们的颓废庸俗。
   2. 石头上可以被切割的地方被标满了记号，反复擦掉又重绘。
   3. 不久后，11月12日，当在摩纳哥参加格雷斯公主的40岁生日聚会时，泰勒小姐第一次公开佩戴"泰勒 - 伯顿"。
   4. 泰勒小姐宣布她正在公开销售该钻石，并计划将部分收益用于在博兹瓦纳建立一所医院。
   5. This rule is not applicable to him at all.
   6. So far, Harry has not commented on these reports.
   7. It yielded a profit of at least $36 million.
   8. Here is your opportunity to acquire a luxurious diamond necklace.

◆ **Text（Translation）**

"泰勒 - 伯顿"钻石和它的历史

钻石不施怜悯……"如果可以，他们会给穿戴者增添光彩。"这是英国著名作家艾丽丝·默多克早期的小说《沙堡》中一个角色所说的话。现在对一些女性来说可能就是这样——通常佩戴大得夸张的珠宝首饰，在一定程度上显露了她们的颓废庸俗——但它适用于伊丽莎白·泰勒吗？她从她的第五个丈夫，已故的理查德·伯顿那里收到的这些高调的礼物无疑提升了她的外在气质，一点儿都不会不合适。宝石与其佩戴者之间建立了兼容性。

到目前为止，在理查德·伯顿购买的钻石中最知名的是69.42克拉梨形的，后来被称为"泰勒 - 伯顿"的钻石。它是由1966年在总理矿区发现的一个重240.80克拉的粗糙石头切割而成，随后由哈里·

温斯顿收购。哈里当时说:"我认为在这个世界上这样质量的石头不会有6块。"

粗糙的石头到达纽约后,哈里·温斯顿和他的切割师 Pastor Colon Jr. 研究了6个月。石头上可以被切割的地方被标满了标记,擦除又重绘。在他切开石头之后,这位50岁的切割师什么也没有说——他穿过工作台,找到了一块分离出来的钻石,然后通过他的仿角质镜架眼镜看了一会儿,惊叹道:"太美了!"这件约78克拉的石料预计能割出一块约24克拉的钻石,而重162克拉的大块石料绝对是为了割出一块预期为约75克拉的梨形钻石而生。

在哈里·温斯顿之后,石头的第一个所有者实际上不是伊丽莎白·泰勒。1967年,温斯顿在理查德·尼克松政府期间把梨形钻石卖给了美国驻伦敦大使的妹妹。两年后,钻石被拍卖。伊丽莎白·泰勒是其中一个名字,她确实参加了这个钻石的预展。

拍卖人开始投标,问是否有人会出价20万美元,在那个拥挤的房间里同时爆发了"有"。招标开始爬升,有9个投标者活跃出价,冲到50万美元。在50万美元,个人出价以1万美元为增量。拍到65万美元,只剩下两个投标人。当投标达到100万美元时,代表理查德·伯顿的弗兰克·波拉克的阿尔·尤格勒退出了。锤子一拍,大家在房间里站了起来。获得者是肯莫尔公司董事会主席罗伯特·肯莫尔,他是卡地亚公司的老板,他以创纪录的价格105万美元买下了这块钻石,并立即将其命名为"卡地亚"。上一个纪录是在1957年从 Rovensky 地产卖出的一条钻石项链,当时价格为30.5万美元。

但伯顿还没有停止,并决心收购钻石。因此,从一个英格兰南部知名酒店的投币式电话中,他与肯莫尔先生的代理人进行谈判。最后罗伯特·肯莫尔同意出售它,但是要满足一个条件,即允许卡地亚公司在纽约和芝加哥展出现今以"泰勒-伯顿"为名的该钻石。

不久后,11月12日,当在摩纳哥参加格雷斯公主的40岁生日聚会时,泰勒小姐第一次公开佩戴"泰勒-伯顿"。从纽约飞往意大利尼斯,它由伯顿和卡地亚雇用的两名武装警卫保护。1978年,在她与理查德·伯顿离婚后,泰勒小姐宣布她正在公开销售钻石,并计划将部分收益用于在博茨瓦纳建立一所医院。1979年6月,纽约珠宝商亨利·兰伯特说,他以500万美元买了"泰勒-伯顿"钻石。

到了12月,他把石头卖给了其现在的持有者罗伯特·穆瓦德。不久之后,穆瓦德先生对其进行了轻微重切,现在重68.09克拉。

Part Ⅲ　Extended Learning

Dictation

1. auction　2. Understanding　3. appraisal　4. valuation　5. is composed of
6. seized　7. confiscated　8. coordinate　9. objective　10. immediately

◆ **Read More**(**Translation**)

Passage 1

现代首饰历史

欧洲文艺复兴对所有的艺术和创新都是有利的,它为随后的宗教改革开辟了道路。我们发现当时伟大的艺术家,如莱昂纳多·达·芬奇都是通过与当时的著名金匠一起工作而开始了他们的职业生涯。众所周知,达·芬奇的整个职业生涯都在为一些支持他的艺术和创作的杰出人士设计珠宝。所有艺术作品的质量达到了新的水平。这个时代的工匠们把更现代的方法带进了整个珠宝贸易上。

欧洲各地上演着各种珠宝争相媲美。皇室成员试图以各种方式相互超越,包括哪个宫廷拥有最好和最奢侈的宝石。这种类型的比拼新颖,达到前所未有的水平。

英格兰的亨利八世对珠宝有着极大的热情,这种热情由他的女儿伊丽莎白一世延续着。据历史记载,小汉斯·霍尔拜因在亨利八世统治时期将文艺复兴饰品引入英国。他是君主个人首饰的主要设计师。亨利制造了数百枚戒指,他特别喜欢镶有宝石的帽子。

这些时期的女性经常佩戴一条以上的项链。这种习惯的特点是先穿上一条紧围颈部的短项链,再戴另一条长项链,比如一条长珍珠项链。有时,还戴上第三件以作为突显物,如在披肩或其他宽松的衣服扣上一个扣子。

吊坠似乎是最受欢迎的珠宝类型，而用于生产它们的技术是精湛的。有些古典的浮雕宝石也很受欢迎。后来添加黄金和其他宝石给予吊坠丰富多彩的外观。彩饰用来生产各种不同主题的珠宝，如使用镂空图案和几个相连的饰品。

在欧洲，戒指有各种各样的风格。某些戒指可以打开和闭合，也会有毒物或其他物质存储在那里。传说中的上流社会人们佩戴的戒指内含毒物，而这正好成为作家编造故事的灵感来源。

在文艺复兴艺术贡献的后期，人们对巴洛克艺术形式开始越来越感兴趣。这种艺术形式在 17 世纪特别流行，趋势遍及欧洲各地。其间，工匠们花时间真正地观察和研究他们的工艺，并改进它。对花卉艺术有一定的倾向。花卉成为高级珠宝设计师的设计主流。其他主题也有呈现，但花卉似乎是最受欢迎的。

在 17 世纪，应用钻石和其他宝石的人越来越多。人们开始佩戴令人惊叹的高品质的珠宝，而不是穿戴一些普通的宝石。佩戴首饰最奢华的君主是法国的路易十四。

巴西成为寻找钻石的宝石猎人的好地方。宝石切割成为整个珠宝贸易的重要组成部分，在很大程度上利于钻石的扩散。

韦奇伍德是一位英国陶艺家，他投身到珠宝贸易行业，并对珠宝生产中涉及的技术做出了重要贡献。他在各类首饰中引入了椭圆形、八角形和圆形的饰板。

19 世纪的工业革命基本上创造了一个对公众开放的珠宝市场。中产阶级现在可以购买精美的首饰，当仿制品开始泛滥时，即使那些来自工人阶级的人也能买到一件首饰。

商业企业随即成立并公开销售珠宝。费伯哥、卡地亚、蒂芙尼和其他大型珠宝公司在工业革命中起步并稳固根基。新艺术风格和抽象的想法，如立体主义与伟大的艺术家，如萨尔瓦多·达利和巴勃罗·毕加索都促进了珠宝贸易的发展。

珠宝生产历史具有全球性，而精心设计出现在亚洲，主要在印度和中国。然而，珠宝生产从来没有达到它在中东、埃及、古欧洲传统的高度。南美洲的工匠和中美洲的部分地区以金银加工而闻名。早期南美和中美洲金匠作品可以在哥伦比亚波哥大的黄金博物馆和墨西哥城的历史博物馆找到。

Passage 2

女王的珠宝

女王的珠宝是一个具有历史价值的收藏品系列，是由英联邦王国的君主私人拥有的珠宝，目前为伊丽莎白二世女王。女王的珠宝与英国皇冠珠宝不一样。人们认为此珠宝起源于 16 世纪但与官方皇冠珠宝不同的皇家珠宝系列的起源是模糊的。许多珠宝来自遥远的国家，由于内战、政变和革命而被带回英国，或者作为礼物赠送给君主。

英联邦官方皇冠仅在加冕（圣爱德华的王冠用于冠君）和国家开放议会（帝国政府冠）时佩戴。在其他正式场合戴着头饰。当女王出国时，她在正式活动中佩戴个人收藏的冕式头饰。

大不列颠和爱尔兰女孩头冠

大不列颠和爱尔兰女孩头冠是英国和爱尔兰女孩在 1893 年赠予未来皇后玛丽的礼物。这顶钻石头冠是由格雷维尔夫人组织的委员会从伦敦珠宝商 Garrard 那里购买的。1947 年，玛丽把头冠作为一个结婚礼物送给她的孙女，未来的伊丽莎白二世。

Leslie Field 描述这个头冠为"钻石花絮和滚动设计，由 9 颗大东方珍珠镶嵌在钻石钉上，并镶嵌在两个简单的钻石之间圆形和菱形夹头交替的缎带底座上"。

伊丽莎白二世通常佩戴没有基座或珍珠的头饰，但近年来，人们发现其已经重新嵌上基座。

多年来，由于其出现在英国的钞票和硬币上，这顶头冠已成为人们最熟悉的伊丽莎白二世女王的皇冠之一。

安德鲁公主希腊回纹头冠

这顶头冠是伊丽莎白公主当时的婆婆希腊和丹麦安德鲁公主（前巴腾堡公主）作为一个结婚礼物赠

予她的。这顶头冠以古希腊风格作为造型设计，中间有一颗硕大的闪耀切割钻石，周围环绕着钻石花环。它还包括一个中央的叶子花环和两边的卷轴。女王从来没有公开佩戴该头冠，并在 1972 年赠予安妮公主。安妮公主经常在公众场合佩戴，特别是在她与菲利普斯订婚时以及她 50 岁生日时官方拍摄的肖像中。2011 年 7 月，皇家公主把这头冠借给了她女儿扎拉·菲利普，准备让她在她与迈克·廷德尔的婚礼上佩戴。

1936 卡地亚光环头冠

这个头冠是约克公爵（后来的国王乔治六世）买给他妻子（后来的伊丽莎白女王母亲）的。三个星期后，他们成为国王和王后。它是一个滚动波浪式的，在一个中心装饰品顶部聚拢式的闪耀钻石头冠。随后，在未来女王伊丽莎白二世 18 岁生日的时候，他们送给了她。

1951 年玛格丽特公主 21 岁生日时收到了波斯绿松石头冠。在此之前，她曾借来卡地亚头冠。玛格丽特公主戴着卡地亚头冠参加了 1953 年伊丽莎白二世女王加冕礼。

女皇伊丽莎白二世后来把头冠给她的女儿安妮公主。之后在 1972 年也把希腊回纹头冠送给她。

这个卡地亚头冠被借给了凯瑟琳·米德尔顿，她在 2011 年 4 月 29 日与威廉王子的婚礼上佩戴了该头冠。

Unit 4　Jewelry Appraisal

Starting Out

◆　**Match Words with Pictures**

1. rock crystal　2. Chrysoberyl cat's eye　3. Catier ruby　4. sapphire　5. emerald　6. tourmaline

◆　**Check Your Knowledge**

	RI (refractive index)	SG (specific gravity)	Dis (dispersion)	Natural (Y/N)	Hardness
diamond	2.417	−	0.044	Y & N	10
spinel	1.718 (+0.017, −0.008)	3.60	0.02	Y	8
tourmaline	1.624 − 1.644	−	−	Y & N	7 − 8
sapphire	1.76 − 1.77	4.0 − 4.1	−	N	9
emerald	1.56 − 1.57	2.67 − 2.9	−	Y	7.5 − 7.75
chalcedony	1.535 − 1.539	2.58 − 2.64	−	Y & N	6 − 7

Part Ⅰ　Communicative Activities

◆　**Task 1　Conversation**

1. c　2. e　3. b　4. a　5. d

Part Ⅱ　Read and Explore

◆　**Check Your Understanding**

Ⅰ. 1. supply and demand; individual subjectivity; type of appraisal

　　2. cut; color; clarity; carat

　　3. commercial cut; superior custom cut

　　4. increase

5. clarity; gem; exact measurements; type of cut; carat size

II. 1. Larger supply with constant or less demand will lead to lower gemstone prices, vice versa.
2. The personal preferences of the appraiser can bear on appraised value of a gemstone.
3. It will increase the valuation for gemstones because they are intended to provide coverage for replacement in a market that has fluctuating valuations.
4. Cut is perhaps the most consistently important C when it comes to determining the value of a gemstone.
5. A gemstone certification will grade the gem's clarity in addition to mapping the gem, and listing the gem's exact measurements, type of cut, and carat size. While a gemstone appraisal is offered after a certification has been issued for the gem.

◆ **Language Focus**

I. 1. scarce   2. accredited   3. saturation   4. evolving   5. fluctuate
   6. locate   7. appraisal   8. mediocre   9. dramatically   10. subjectivity

II. 1. ballpark   2. preference   3. wash out   4. consistently   5. evaluation
    6. appraise   7. overall   8. intensity   9. reverse   10. faded

III. 1. Even 改为 Even if   2. 删去 but
     3. 删去 Although 或 but   4. While 改为 Although/Though
     5. 删去 but   6. But 改为 However
     7. nevertheless 改为 but nevertheless 8. 删去 but 或 whereas
     9. In spite 改为 In spite of   10. 删去 of

IV. 1. 宝石证书将界定宝石的净度，标注该宝石的产地，列出宝石精确尺寸、切割类型和克拉大小。
2. 虽然（珠宝）随着时间推移而增值，但任何特定珠宝的估价都可能会有波动。
3. 饱和度是宝石颜色的强度和深度，可洗掉的或褪色的宝石比饱和度高的宝石的需求度低得多。
4. 在某种程度上，鉴定者的个人偏好会影响某一珠宝的估值。
5. The color, luster and clarity of a gemstone can first be noted with the unaided eye in gemstone appraisal.
6. Refractometers and scales are important instrument in gemstone appraisal.
7. We need to identify not only the types of a gemstone, but also its formation—whether it's natural or synthetic, whether it's been artificially optimized or treated.
8. When it comes to determining a gem's value, the importance of each "C" can change dramatically depending on the type of gemstone being evaluated.

◆ **Text（Translation）**

宝石鉴定及宝石估价

宝石会随着时间而增值，尤其是高品质的宝石，因其稀少，需求量大而更是如此。

虽然珠宝会随着时间推移而增值，但任何特定珠宝的估价都可能会有所波动。很大程度上，这种波动基于以下三个原因：供求关系、个人的主观性和鉴定类型。

供求关系

供和求都会引起珠宝价值的波动。举例来说，供方价值增长可能是由于地理资源有限，比如坦桑黝帘石和沙弗莱石。如果供应中断（比如地方或矿井泄洪）而需求保持不变，那么这类珠宝的价值肯定会大幅增长。反之亦然，例如原本供应量稀缺的某种宝石因突然发现新的矿井而供应量充足，则这种珠宝的价格就会降低。

个人的主观性

在某种程度上，鉴定者的个人偏好会影响到某一珠宝的估值。切割、颜色、净度和克拉重量从根本上决定了某一珠宝种类的需求度，而这四个变量孰轻孰重则会受到鉴定者的个人喜好的影响。比如，对某个鉴定者来说，纯绿色的祖母绿比有点蓝绿色的祖母绿级别更高更贵，但对另一个鉴定者来说，情况则相反。

鉴定类型

重置成本评估，一般称为保险评估，被用于资产损失时对个人的赔偿，往往会导致宝石和珠宝的估值过高，因为该评估旨在估值波动的市场对重置进行赔偿。

宝石鉴定是一个不断发展的科学，珠宝学家们也在不断确定哪些标准更适用于珠宝鉴定。各种不同的宝石的颜色、形状和大小都增加了鉴定过程的复杂性。

目前没有行业标准适用于珠宝估价（除了钻石），但珠宝学家们还是可以基于珠宝的4C：切工、颜色、净度和克拉重量，来对珠宝进行大致正确的估价。

当然，在决定珠宝价值时，每个C的重要性会因所鉴定的珠宝种类而不同。

下面，你将发现关于珠宝等级、珠宝证书、珠宝鉴定的一些有用信息，这些信息你在购买珠宝和珠宝用品时会用得上。

彩色宝石中切工的重要性

在确定宝石价值时，切工可能一直是最重要的C。切割不到位或一般性商业切割都没有可能显示出宝石的光彩夺目。高级的定制切割可以区分好的宝石和令人惊叹的宝石。

顾客很难即时区分出商业切割和高级定制切割，但多花点时间去观察同一宝石的不同的切割会让顾客清楚高质量切割和普通切割之间的差异。

宝石估价的其他 3 个 C

颜色在不同的宝石种类的估价中有举足轻重的作用。以蓝宝石为例，淡黄色的蓝宝石均价在每克拉500元左右，稀少的但需求量大的、落日红或橙红色的巴特帕拉德石蓝宝石均价在每克拉5000元到15000元之间。

宝石颜色的饱和度对于宝石鉴定也有影响。饱和度是宝石颜色的强度和深度，可洗掉的或褪色的宝石比饱和度高的宝石的需求度低得多。

净度也会影响宝石的价值。像翡翠这类宝石是由包含物天然形成的，只要肉眼看不见这些包含物，它们就不会影响宝石的整体价值。但其他类型的宝石，诸如海蓝宝石，净度是决定宝石价值的一个重要因素。

克拉大小对宝石价值的影响取决于被评估的宝石类型。上面提及的海蓝宝石是一个典型例子，这类宝石往往克拉数很大，因此克拉大小对海蓝宝石的估价没有什么影响（一个40克拉的海蓝宝石和一个5克拉的海蓝宝石的价值可能一样）。其他像蓝锥矿之类的宝石，因为很稀少，大克拉的蓝锥矿更少见，因此它们的价值很大程度上取决于克拉大小。

宝石鉴定和宝石证书

我们要明白宝石鉴定和宝石证书是两种不同的概念。

宝石证书将界定宝石的净度，标注该宝石的产地，列出宝石的精确尺寸、切割类型和克拉大小。

而宝石鉴定往往在给宝石颁发证书之后提供。许多独立的宝石实验室只提供证书，但官方认可的独立珠宝鉴定学家可以为宝石提供证书和鉴定。

Part Ⅲ Extended Learning

◆ **Dictation**

1. hurry up 2. delegates 3. market circulation 4. porcelains 5. at the scene
6. agents 7. auction season 8. promising 9. turnover 10. abundant

Key to the Exercises and Translation

◆ **Read More**（Translation）

Passage 1

宝石鉴定提示

当你观察相似宝石时，一些提示可以帮你鉴别。

如果宝石呈现出各种光学现象，比如变彩效应、变色效应和晕彩效应，宝石的可能性品种会大大减少。我们在一些宝石品种中会发现弱的星光效应或猫眼效应。星光效应常见于红宝石、蓝宝石和正长石中。能显示猫眼效应的宝石包括常见的金绿宝石、石英、碧玺，也包括绿柱石、翠榴石、软玉、顽火辉石、透辉石、长石、磷灰石、锆石、硅线石和其他宝石。

将一个透明的刻面宝石台面朝下放置在白色背景下时，其腰棱翻光面附近出现红环（圈）效应，说明它是以石榴石为顶层的拼合石。从深处看，闪红光、具有艳蓝色的宝石可能是合成尖晶石，或者是黝帘石的亚种坦桑石。

锆石和合成金红石都有极高的双折射率，放大观察时很容易从透明宝石种类中将其辨认出来。

如果宝石含有天然包裹体，又有很强的后刻面棱重影，是合成金红石的可能性可以被排除，那么未知物必定是锆石。可以通过放大观察双折射宝石，或者，如果有气泡又不具双折射，则是玻璃。但是，如果宝石被证明是钻石，则只有分光镜可以区分是天然的还是人工着色的。

如果宝石具有双折射但没见包裹体，则可依据相对密度、双折射的强弱和色散值的高低区分高型锆石和合成金红石。根据宝石轮廓突起的巨大差异，浸没在二碘甲烷中的宝石可显示明显不同的折射率值。

Passage 2

钻石——岂止于美

作为"女性的好朋友"，钻石因耳环、头冠和《粉红豹》之类的电影而闻名。但这些形成于数百万年以前的地壳深处，经历极度的高压高温，光彩四射的珠宝的审美用途却不多。今天，从外科医生的解剖刀到极快的微芯片，许多钻石因用途广泛而人工合成。

实验室每年生产180吨钻石——几乎是天然钻石产量的9倍。钻石的强度、净度和化学耐力使它成为21世纪的工程材料。目前已经有小型合成电路的钻石加热槽，接合处的钻石涂料和空间探测器的钻石窗。很快就会有钻石涂料来保护你汽车的变速箱，超坚固的钻石线用来加固超轻飞机。

钻石不仅仅是漂亮的石头。它还有一系列令人印象深刻的特点。作为科学界已知最硬的材料，它可以抵制强酸强碱的侵袭，而且导热性能极好。这一切意味着除了纯装饰性，钻石还有其他很多用途。因为它散热性如此之好——远远好于硅——以至于工程师们想用钻石层来做微芯片，这样，避免电路过热，他们就可以把更多的电子成分压进更小的区域来制造新一代的超快计算机。钻石特质的关键在于它的结构。碳可以和其他分子形成四种牢固结合，这也是它可以形成这么多有机化合物的基础，是构筑生命的重要单元。当四个碳原子连接成一个规则的格子，就形成了钻石晶体。碳的另一种形式是铅笔中的石墨。

天然钻石形成于距今300万年的地壳以下200千米的岩浆覆盖物中，然后它们混在火成岩，如金伯利岩（含金刚石）中被向上带。当融化的岩石到达地球表面时，会冷却形成管状结构，而天然钻石就是在这些管状结构中发现的。

戴比尔斯和通用电气这类公司自20世纪50年代早期已经在生产合成钻石了。几乎任何富含碳的物质都可以转化为钻石。通用电气的化学专家罗伯特·文托夫曾经用花生酱制作出钻石。

一个更新的工艺——化学气相沉积（CVD），用于生产超硬的钻石涂料。CVD在高温低压下用碳蒸汽覆盖在物质表面，就像一层小的钻石晶体。这些晶体最终会结合在一起，被用来制造大的宝石。这类钻石往往被切片用于生产长的手术刀或其他工具。

CVD让科学家为之振奋。通过表层覆盖钨线用CVD来生产钻石线的玫说："这是我们首次将钻石所有最高级的特点以特定的形式应用于工程。"所以，钻石似乎不仅仅是女孩子的好朋友，也是工程师的好朋友呢。

149

Unit 5 Diamonds

Starting Out

◆ **Match Words with Pictures**

1. Great Star of Africa Ⅰ 2. Great Star of Africa Ⅱ 3. Great Mogul 4. Kohinur
5. Star of Sierra Leone 6. Golden Jubilee Diamond 7. Weyie River Diamond 8. Eureke

◆ **Check Your Knowledge**

1. C (carbon) 2. Cubic 3. Diamond is the hardest known natural substance. The hardness of diamond varies according to the crystallographic orientation. 4. 3.52 g/cm^3 5. Colorless; yellowish, brownish or greenish. Fancy colors (those of a distinct hue) include yellow and brown, rarely green, pink and blue, very rarely red and purple. 6. 2.42 single. Many diamonds display anomalous extinction. 7. High 0.044. Diamond displays a higher degree of dispersion than any other natural colorless gemstone.

Part Ⅰ Communicative Activities

1. a 2. c 3. b 4. d 5. e

Part Ⅱ Read and Explore

◆ **Check Your Understanding**

Ⅰ. 1. carat; clarity; color; cut
   2. A carat
   3. LC; VVS; VS; SI; P
   4. CZ

Ⅱ. 1. It is 10.
   2. It means colorless, yellowish, brownish or greenish. Fancy colors (those of a distinct hue) include yellow and brown, rarely green, pink and blue, very rarely red and purple.
   3. Important gem diamond producing countries include Angola, Australia, Botswana, Brazil, China, Namibia, Russia, Sierra Leone, South Africa, Tanzania.
   4. Many natural, synthetic and artificial products have been used to imitate diamonds.
   5. CZ, Cubic zirconia, is by far the best and most widely used simulant of diamond today.

◆ **Language Focus**

Ⅰ. 1. platinum   2. exquisite   3. indented   4. blemish   5. refractive index
   6. macle   7. pits-trigons   8. luminescence   9. phosphorescence   10. deposit

Ⅱ. 1. ensure   2. distorted   3. decrease   4. appraise   5. trim off
   6. by far   7. extinction   8. grade   9. In addition to   10. in terms of

Ⅲ. 1. was 改为 is   2. was 改为 were   3. been 改为 be   4. 删去 been
   5. being 改为 be   6. be 改为 being   7. haven't 改为 hadn't   8. exciting 改为 excited
   9. dressing 改为 dressed   10. were 改为 are

Ⅳ.
1. 在珠宝贸易中，钻石是所有宝石中最重要的种类。据估算，钻石销售占全球珠宝贸易额约90%。
2. 一般而言，钻石分为工业级和宝石级两种，后者通常用"4C"分级标准：克拉重量、净度、颜色和切工。
3. 钻石的荧光变化可用于鉴别群镶的无色钻石。
4. 许多天然的、合成的及人造品被用于仿钻石。其中外观最相近的有：①合成立方氧化锆（CZ）；②钇铝石榴石（YAG）；③无色锆石；④某些玻璃。
5. Except for stones of fancy color, the desirability and value of diamond decreases as the depth of hue increases.
6. The fluorescence of diamond varies in color and intensity.

7. If all stones set in a piece show similar fluorescence, the stones are unlikely to be diamond.
8. CZ is by far the best and most widely used simulant today.

◆ Text（Translation）

钻石的宝石学特征

在珠宝贸易中，钻石是所有宝石中最重要的类别。据估算，钻石销售占全球珠宝贸易额约90%。钻石总是被加工成小刻面以展示其独一无二的光泽和火彩。它至高的硬度确保其切工的精度，也是所有宝石中唯一的。

一般而言，钻石分为工业级和宝石级两种，后者通常用"4C"分级标准：克拉重量、净度、颜色和切工。除了标准圆钻形外，钻石也常被加工成梨形、椭圆形、心形、马眼形、三角形或祖母绿正方啄形。通过精心的设计，钻石可被镶嵌到所有精美的首饰中，如项链、耳环、戒指等。钻石的宝石学特征如下：

（1）化学成分：C（碳）。

（2）晶系：等轴晶系。

（3）晶体习性：最重要的是八面体。钻石也呈立方体、菱形十二面体、似立方体等。钻石晶体经常畸变，晶面可能弯曲。可有八面体晶体（双晶常见）。

（4）表面特征：可见八面体晶面上的三角形蚀象。解理：完全的八面体解理，可用于钻石抛光、开料或修整有瑕疵的钻石材料，可见于成品钻石和钻坯的内部及表面。

（5）莫氏硬度：莫氏硬度为10的钻石是自然界已知物质中最硬的。钻石的硬度依结晶方向不同而改变。但在任何方向，钻石依然比其他宝石硬得多。

（6）比重：3.52克/立方厘米。

（7）颜色：无色、浅黄色、褐色或绿色。彩色（具有明显色调的那些）包括黄色和褐色，少见有绿色、粉色和蓝色，非常罕见的是红色和紫色。除了彩色钻石外，钻石的价值随着色调的加深而降低。

（8）光泽：金刚光泽。

（9）折射率：2.42，单折射。许多钻石具有异常消光。

（10）色散值：0.044（高），钻石的色散值比其他任何天然无色宝石都高。

（11）发光性：钻石的荧光变化范围较大，从无到强，荧光颜色多变。长波紫外光下的荧光强度比短波紫外光下的强，为蓝白色到紫色、绿色或黄色荧光。也有些钻石呈现荧光惰性。钻石的荧光变化可用于鉴别群镶的无色钻石。如果一件首饰上所有的宝石都呈现相似的荧光特征，则这些宝石就不太可能是钻石。

紫外灯下呈蓝色荧光的那些钻石可能具有黄色磷光，这是鉴别钻石的诊断性依据；

（12）产状：主要产于金伯利岩管或冲积矿床中。

（13）产地：印度的冲积矿床是古典时代到18世纪间已知的唯一钻石矿资源。巴西的重要钻石矿藏大概发现于1725年。

在19世纪后半叶发现了南非冲积矿床和金伯利岩管钻石矿床，西伯利亚钻石矿床发现于20世纪40年代。近来，澳大利亚成了重要的钻石产地，产于与金伯利岩相似的钾镁煌斑岩中。

宝石级钻石的重要生产国有安哥拉、澳大利亚、博茨瓦纳、巴西、中国、纳米比亚、俄罗斯、塞拉里昂、南非和坦桑尼亚等。

（14）相似宝石：许多天然的、合成的及人造品被用于仿钻石。其中外观最相近的有：①合成立方氧化锆（CZ）；②钇铝石榴石（YAG）；③无色锆石；④某些玻璃。

到目前为止，合成立方氧化锆是最好的和应用最广泛的仿钻品。其他用作仿钻的有天然和合成的白色蓝宝石、合成白色尖晶石。

English for Jewelry

Part Ⅲ   Extended Learning

◆ Dictation

1. good investment
2. 160 percent/160%
3. unchanged
4. diamond prices
5. depreciated against
6. increasingly shaky
7. deposit interest rates
8. supply chain
9. retail stores
10. diamonds' 4C criteria

◆ Read More (Translation)

Passage 1

如何清洁钻石首饰

　　珠宝或首饰，如项链、手链和耳环等会提升一个人的美丽和魅力。尽管大家都喜欢耀眼的珠宝，但主要是女性对它钟情。佩戴数日后，那明亮和闪光的珠宝可能就会失去它的魅力和灿烂。你想知道为什么发生这种情况吗？这是由于灰尘、污垢和我们使用的护肤品。珠宝商经常建议我们每天仔细清洁珠宝。认真清洁可以把一件失色的珠宝变得迷人和引人注目。

　　清洗珠宝之前，你应该确保此类事项，如宝石应该正确地安装。确保所有的钩扣、分离双环等把整体固定在一起的部分都足够牢固以备清洁。如果你不确定如何清洁你精致的珠宝，那么自己在家清洁之前，跟你的珠宝商谈谈是很重要的。现在，我们想到的问题是如何清洁珠宝，要回答这个问题，我可以告诉你清洁珠宝有各式各样的方法，自己在家也可以清洗。

　　清洁银首饰小贴士：一件银首饰应该用软棉布或法兰绒布清洁。

　　＊银质清洁布可以帮助快速清洗银首饰，因为它有防锈成分。

　　＊一支柔顺的牙刷可以用来清洁首饰错综复杂的漩涡形装饰。

　　＊在温水中混合少量的液体洗涤剂或肥皂也可以用来清洁污垢。

　　＊用温水洗净首饰，然后用布擦干、晾干。

　　＊如果你有使用牙刷，轻轻擦洗，冲净。

　　＊使用银渍剂（液体清洁剂）或小苏打膏去除污点。

　　然而最好避免使用银渍剂清洁宝石，因为宝石由于化学反应可能会受损。

　　清洗黄金首饰小贴士：清洗一件迷人的黄金首饰，如项链、戒指、镶嵌着美丽宝石的手镯，则要浸泡在热肥皂水中10到15分钟。然后用软毛牙刷刷洗，在温水中冲净，然后晾干。再加几滴氨水到肥皂水中去除污点。此外，通过浸入酒精，污垢可以从你的金饰中去除。假设清洁白金首饰，可以在珠宝商协助下使用超声波清洗珠宝。

　　清洗钻石首饰小贴士：每个女孩都有对佩戴迷人的、珍贵的钻石珠宝的狂热渴望。像其他首饰一样，它也可能在频繁佩戴后失去其光芒。要让你的钻石珠宝重新回复光彩，你应该知道如何清洁小部件。以下方法可以帮助清理你的钻石首饰：

　　＊把你的钻石耳环、项链等浸泡在温水和氨的混合溶液或温和的清洁剂中。

　　＊现在用牙刷轻柔地刷洗。

　　＊可用牙签处理够不到的地方。

　　＊让它在干净的布上或者纸巾上晾干。

　　在清洁你的钻石首饰的时候，不使用任何自制的含氯漂白剂，因为这会破坏首饰的美丽和闪耀。

　　清洁珍珠小贴士：清洁由精致珍珠制成的珠宝，你要用柔软干净的布，并在肥皂水中浸湿。接着用这块湿布擦拭。轻轻冲洗它，再晾干。不要把珍珠首饰浸泡在肥皂水中，因为它可能会导致伸展。一件漂亮的珍珠项链、耳环、手镯等应在化妆和擦香水后佩戴。

　　这些珠宝清洁技巧可以帮助你心爱的首饰重获光彩。当清洁珍珠首饰时，建议不使用磨料或其他含酒精的混合物，因为酒精可能会损害首饰。为达到更好的效果或专业清洁，你可以找一个专业的珠宝人士去擦亮你的项链或手镯。

◆ **Passage 2**

4C 标准等级鉴定

钻石颜色等级鉴定

因光源和背景对钻石外观有非常大的影响，故可在标准化的观察环境中鉴定钻石颜色等级。可配备至少两名颜色等级鉴定师分别为同一颗钻石做鉴定。对于鉴定结果不一致的，或是一些重量和品质特殊的钻石，会有更多鉴定师进行颜色等级检测。直到检测结果达到高度一致，才会最终确定钻石的等级。

钻石净度等级鉴定

钻石净度等级鉴定采用 10 倍放大镜在标准观测条件下进行。进行初检的等级鉴定师会仔细检查钻石，确定钻石的净度或涂饰抛光特征，并采集如裂隙充填或激光钻孔等净度处理的证据。

至少两名等级鉴定师将对钻石的净度、打磨程度和对称性进行鉴定，还会将决定净度的内含物特征绘制到显示钻石的形状和刻面形式的图上。

钻石切工等级鉴定

GIA 为颜色等级在 D（无色）至 Z（淡黄或淡褐色）之间的标准圆形钻石提供切工等级鉴定。为了确立切工等级鉴定系统，GIA 进行了为期 15 年的针对圆形钻石的广泛计算机建模，并对天然钻石进行了 70 000 多次的观察，以核实其研究成果。目前，该系统可以凭借其超过 3850 万组的比例预设值，对钻石进行切工等级预测。

GIA 的钻石切工等级涵盖从极优至不良，用于鉴定钻石正面朝上的整体外观，从而预测其明亮度、火彩和闪光（钻石与光线互动产生的光芒）等级。GIA 也会查验每一颗送检的钻石，确定其是否为合成钻石。

钻石克拉重量测定

GIA 使用非常精确的电子微量天平进行称重，测定钻石的克拉重量，电子微量天平可精确到小数点后 5 位（即一克拉的万分之一）。采用光学测量装置来确定钻石的比例、尺寸和刻面角度。

Unit 6　Ruby, Sapphire and Beryl (Emerald)

Starting Out

◆ **Match Words with Pictures**

1. amethyst　2. aquamarine　3. beryl　4. star ruby　5. padparadscha sapphire　6. garnet

◆ **Check Your Knowledge**

1. faceted cut　2. plain cut　3. mixed cuts

Part Ⅰ　Communicative Activities

1. a　2. c　3. d　4. b

Part Ⅱ　Read and Explore

◆ **Check Your Understanding**

Ⅰ.

	Emerald	Ruby	Sapphire
Birthstone	May	July	September
Color	green	red	violet-blue
Symbol	love and rebirth	love, passion, courage and emotion	honesty, loyalty, purity and trust
Status	the gem of Venus	the king of all gems	one of the most popular engagement gemstones

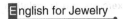

续上表

	Emerald	Ruby	Sapphire
Important features	inclusions	the wide range of red hues	come in almost every color of the rainbow

Ⅱ. 1. Egypt. Because one of their female monarch was very famous for wearing emeralds and ancient Egyptian mummies were often buried with an emerald.

2. Because among the rarest of gems, emeralds are almost always found with birthmarks, which is known as inclusions.

3. Because it was believed that wearing a fine red Ruby bestowed good fortune on its owner. Rubies have been the prized possession of emperors and kings throughout the ages.

4. "Pigeon blood" is associated with the color of a white pigeon's eye.

5. According to the text, the deeper and more vivid the color of green, the more valuable the emeralds. The brightest and most valuable color of Ruby is often "a Burmese Ruby", or "pigeon blood" Ruby. The purer the blue of the Sapphire, the greater the price.

◆ **Language Focus**

Ⅰ. 1. inclusion 2. hue 3. saturation

Ⅱ. 1. indicates 2. ultimate 3. bestowed 4. symbolize 5. stunning
6. envision 7. eternally 8. reminiscent 9. intense 10. distinctive

Ⅲ. 1. He knew his behavior was childish.
2. She had reddish brown hair.
3. Dan is a smart, dark-haired youngish man.
4. She gave a little girlish giggle.
5. It is commonly believed that people tend to be selfish.

Ⅳ. 1. 宝石的颜色种类有如彩虹般绚丽多彩，遍布世界各个角落，每颗彩色宝石都拥有一种独特的色彩。

2. 埃及的女暴君克利奥帕特拉在位期间以喜欢佩戴翡翠著称，正如伊丽莎白泰勒在我们这个时代出了名喜欢钻石一样。

3. 翡翠的颜色越深，越生动，其价值就越高。

4. 自古以来，蓝宝石就代表着对诚实、忠诚、纯洁和信任的承诺。

5. This situation will inevitably play a role in facilitating the formation of sub-markets.

6. Green symbolizes love and rebirth.

7. Some inclusions are expected in emeralds and do not detract from the value of the stone.

8. Rubies are available in a range of red hues from purplish and bluish red to orange-red.

Part Ⅲ Extended Learning

◆ **Dictation**

1. marketplace 2. run through 3. sponsored 4. exhibitors 5. indicates
6. raw materials 7. gemstones 8. renowned 9. craftsmanship 10. trends

◆ **Text（Translation）**

宝石指南

纵观历史，宝石在人类文化各阶段的神话和传说中发挥了各种作用。有些宝石就是一个故事，人们相信它们具有特殊力量，但所有宝石都同样美丽。每颗宝石都是独特的，具有特别的颜色、出生地和故事。宝石的颜色种类有如彩虹般绚丽多彩，遍布世界各个角落，每颗彩色宝石都拥有一种独特的色彩。

一些宝石自有历史以来就被收藏，而其他一些宝石最近才被发现。跟我们一起来探索彩色宝石世界吧。

翡翠：五月诞生石

绿色是春天的颜色，一直以来象征着爱与重生。作为金星的宝石，它也被认为有助于生育。

埃及的女暴君克利奥帕特拉在位期间以喜欢佩戴翡翠著称，正如伊丽莎白泰勒在我们这个时代出了名喜欢钻石。古埃及的木乃伊通常用翡翠项圈陪葬，象征着永葆青春。

翡翠的颜色越深，越生动，其价值就越高。最宝贵和美丽的翡翠除了基本的青翠的绿色外，还呈现出强烈的蓝色色调。在最常见的宝石中，祖母绿被发现之时总是伴随胎记。这种胎记被称为内含物。祖母绿一般都会存在内含物，而这并不会像其他宝石一样损害其价值。

红宝石：七月诞生石

红宝石代表爱、激情、勇气和情感。几个世纪以来，这颗宝石被认为是宝石之王。人们认为，佩戴一颗精致的红宝石能赋予它的主人好运。红宝石一直是皇帝和国王珍贵的财产。时至今日，红宝石依然是最宝贵的宝石。

红宝石的颜色是它最重要的特征。红宝石色调涵盖了从紫红色和蓝红色到橙红色的红色色调。红宝石中最亮最有价值的颜色通常是"缅甸红宝石"——这表明它是一种丰富且充满激情、热度且带着微蓝色调的饱满红色。这种颜色通常被称为"鸽血"红色，红宝石色只与缅甸的 Mogok 谷矿相关。称为鸽血宝石红的颜色，与鸽子的血色无关，而是白鸽眼睛的颜色。

蓝宝石：九月诞生石

当听到蓝宝石这个词时，许多人立刻想象到一颗令人惊叹的紫蓝色宝石，因为"蓝宝石"这个词是希腊语的蓝色。几个世纪以来，蓝宝石被称为蓝色宝石中的极品。自古以来，蓝宝石就代表着对诚实、忠诚、纯洁和信任的承诺。为了延续这种传统，蓝宝石是当今最受欢迎的订婚宝石之一。

在世界上许多地方都有发现蓝宝石，但最珍贵的蓝宝石来自缅甸、克什米尔和斯里兰卡。具有高度饱和的紫蓝色和"天鹅绒般的"或"令人昏昏欲睡的"透明度的蓝宝石更罕见。蓝宝石的蓝色越纯净，价格越高。然而，许多人发现，较暗色调的蓝宝石同样具有吸引力。

蓝宝石不仅是蓝色的，它们几乎呈现了彩虹的每种颜色：粉红色、黄色、橙色、桃红色和紫色。最抢手的花式颜色蓝宝石是罕见、美丽的巴特帕拉德蓝宝石：介乎粉红色与橙色之间独特的鲑鱼颜色，让人联想到热带的太阳。

◆ **Read More**（Translation）

Passage 1

关于祖母绿的二十件趣事

1. 翡翠是四大名贵宝石之一。其余的是红宝石、蓝宝石和钻石。
2. 高级翡翠比钻石价值更高，因为没有瑕疵的翡翠是非常罕见的。
3. 1 克拉的翡翠比 1 克拉的钻石大，因为翡翠的密度较低。
4. 大多数翡翠都有某种的内含物或瑕疵。比起"瑕疵"，经销商更喜欢把翡翠包含物称作内"花园"（法语）。
5. 大多数翡翠通常用油来填充裂缝并且能够帮助预防意外破损或开裂。
6. 由于翡翠有内含物，在超声波清洁器中清洁这些宝石是不明智的。而应使用温水轻轻地手洗。
7. 翡翠是由绿宝石制成的，就像海蓝宝石一样，但因含有极少量的铬和/或钒而致其变为绿色。
8. 颜色、净度、切工和克拉重量是用于确定翡翠的价值的四个因素。这四个中最重要的是颜色。最好的颜色是生动的绿色或蓝绿色，色彩饱和，没有颜色分区。同样重要的是，翡翠是非常通透的，不会太暗或太亮。
9. 最古老的翡翠约有 29.7 亿岁。
10. 第一块已知的翡翠在大约公元前 1500 年的埃及开采出来。

11. 克里奥帕特拉最喜欢的宝石之一是翡翠。
12. 在16世纪，西班牙人在南美洲发现了翡翠。在此发现之前，印加人已经在使用。
13. 然后西班牙人将这些翡翠作为贵金属在欧洲和亚洲进行买卖，开创了世界上的翡翠贸易。
14. 根据古代民俗，在舌头下放一个翡翠将有助于预见未来。
15. 今天，翡翠产量最多的是哥伦比亚，贡献了全世界50%以上的翡翠产量。
16. 德文郡公爵翡翠是最大的未切割翡翠之一，重1383.93克拉。
17. 合成蓝宝石和红宝石始于1907年，但直到1935年才出现合成翡翠。它是由美国化学家Carroll Chatham成功合成的第一个1克拉的Chatham翡翠，现在在史密森学会展出。
18. 1997年，在北美Yukon地区首次发现了翡翠。但在美国和更北部的地区中，大型翡翠矿床是非常罕见的。
19. 宝石学家通过使用10倍放大镜来判断钻石的净度。翡翠的净度通常用肉眼评估。
20. 伊丽莎白·泰勒拥有的翡翠吊坠项链在2011年以650万美元的价格售出，分解成每克拉约28万美元。

Passage 2

购买宝石的提示

历史上，人们一直在追寻和收藏各种宝石。它们在数千年前的废墟中被发现。它们被珍视为象征着爱的礼物。

一般来说，任何宝石的价格取决于：尺寸、切工、优质的颜色/净度/处理和类型。以下是考虑宝石质量时该询问的一些问题：

- 是否被人工处理过？（参考下文列出的处理方法）
- 是天然矿石还是合成矿石？
- 是否存在显著的划痕、破损或内含物？
- 颜色是否均匀？
- 颜色有多亮丽？（是否鲜艳？）
- 如果你买的宝石是要镶在耳环和袖扣上的，它们是否匹配？

经销商有许多处理宝石的手段。精明的买家询问很多问题，并希望测试效果。这里列出几种处理方法：

- 激光照射：通常海蓝宝石、伦敦蓝黄玉、翡翠和钻石以及其他石头都会经过激光照射处理。这种处理使得颜色更光亮并能清除瑕疵。许多经销商都了解他们卖的石头是否被照射处理过。诚实的人会告诉你他们是否知道做过处理。
- 热处理：紫晶、海蓝宝石、红宝石、坦桑石和黄玉通常在高温下加热以增强颜色。
- 染色：这是最常用的处理方法。在清澈的石头上，比石头的其余部分更暗的裂缝可见染色。有时，染料看起来像擦掉的残留物或白色斑点。青金石和玫瑰石英普遍会经过染色处理。紫水晶和黄水晶也是如此。黑玛瑙在正常处理过程中都会经过永久染色。
- 涂层：碧玉通常浸在石油产品中使得颜色更鲜亮并加以密封。翡翠一般上油；绿松石上蜡。

购买珠子的提示：

- 足够大的孔（这样可以使用更粗的线）。
- 均匀的珠（适合为主）。
- 珠子是否以16英寸长度的线出售——可能的话不要14英寸或15英寸。
- 寻找最好质量的宝石（如果买正品宝石）。
- 确保珠子没有裂开或因孔破裂，因为这会撕断线。
- 颜色鲜艳饱满（则可匹配项链和耳环）。

Key to the Exercises and Translation

Unit 7 Polycrystalline Gemstones

Starting Out

◆ **Match Words with Pictures**

1. quartzite 2. turquoise 3. nephrite 4. Dushan jade 5. serpentine jade 6. jadeite

◆ **Check Your Knowledge**

Type Item	Nephrite	Dushan jade	Turquoise	Serpentine jade	Quartzite
Chemical composition	Calcium magnesium silicate	Calcium aluminosilicate	Copper and aluminum salt	magnesium silicate	Quartz
Hardness	6 – 6.5	4.5 – 5.5	5 – 6	2.5 – 6	7
Luster	Vitreous to greasy	Vitreous to greasy	Vitreous to greasy	Waxy to vitreous	Vitreous
Transparency	Transparent, translucent to opaque	Slightly transparent to translucent	Opaque	Translucent to transparent	Slightly transparent to translucent
Specific Gravity (SG)	2.90 – 3.10	3.25 – 3.36	2.6 – 2.9	2.57	2.65
Refractive index (RI)	1.60 – 1.61	1.56 – 1.70	1.61 – 1.65	1.56 – 1.57	1.54
Colors	White, yellow, green, pink, black	Red, green, white, purple, yellow, black	Green, blue, yellow	Light yellow, lime-green, green, colourless	White, green
Localities	Xinjiang, Qinghai, Russia	Henan Province, Xinjiang	China, Iran, Egypt, USA, Mexico	China, New Zealand, USA, North Korea	China, Burma, India, South Africa

Part I Communicative Activities

1. c 2. f 3. a 4. e 5. d 6. b

Part II Read and Explore

◆ **Check Your Understanding**

I.

Item Tested	Testing Means	Result of Test
Thermal conductivity	the thermal conductivity test	jadeite
Crystallographic structure	a polariscope	polycrystalline
Color	a Chelsea filter	no color treatment
Luminance	forensic level UV lighting	the bangle doesn't release any ultraviolet light, which is a sign of natural grade A jadeite

续上表

Item Tested	Testing Means	Result of Test
Internal structure	magnification	the structure and characteristics of jadeite
Hardness	a hardness test	the hardness of jadeite
Absorption spectrum	a basic spectrometer	natural grade A jadeite
Craftsmanship		of high quality
Conclusion	The bangle is jadeite and is possibly grade A.	

Ⅱ. 1. Because to guarantee it as grade A jadeite requires further tests with more advanced equipment.
2. They are cautious and responsible.
3. (Open.)
4. Jadeite is slightly harder than nephrite and is often used for jewelry. Jadeite is more popular and generally much more valuable by weight than nephrite.
5. The jade market is flourishing with the prices skyrocketing in the past decades.

◆ **Language Focus**

Ⅰ. 1. translucency 2. provenance 3. dimension 4. polycrystalline 5. treatment
6. magnification 7. craftsmanship 8. polish 9. inclusion 10. fracture

Ⅱ. 1. separate 2. consists of 3. Remarkable 4. skyrocketed 5. on behalf of
6. reserve 7. genuine 8. released 9. be subjected to 10. measures

Ⅲ. 1. which 改为 that 2. Which 改为 As 3. who 改为 whom 4. whom 改为 whose
5. that 改为 which 6. which 后加上 were 7. that 改为 whose 8. as 改为 which
9. which 改为 when 10. them 改为 which

Ⅳ. 1. 在紫外线照射下该玉镯并未发光，显示其为 A 级天然翡翠。
2. 吸收光谱所处区域代表其为 A 级天然翡翠。
3. 如要最终检测该翡翠是否经过极其复杂的染色或处理，需要用到更加昂贵的仪器设备，如拉曼光谱仪和傅里叶变换红外光谱仪。
4. 软玉早在公元前 5000 年便出现在中国，而翡翠在中国最早的记录出现在 18 世纪中期大量翡翠原石进口到中国之后。
5. The traditional jadeite jade identification methods mainly relied on refractometer, gemological microscope, Chelsea color filter and spectroscope.
6. Gem quality jadeite jade occurs as a polycrystalline rock and is one of the two materials known as jade.
7. A combination of great toughness with a subtle beauty makes jadeite jade ideally suited to carving, or to fashioning as beads and cabochons.
8. Nephrite is a non-renewable resource, and quality raw nephrite resources are very rare.

◆ **Text（Translation）**

中国玉镯：白色半透明翡翠

白色翡翠玉镯

半圆柱形。

白色，高硬度，半透明质地。

经检测为天然翡翠，下附鉴定结果。

鉴定结果后附玉镯的玉龄、性状和尺寸。

我们保留取消该商品目录的权利。

更多信息可参见销售信息。

我们的专家已经对该玉镯进行了以下方面的检测：

1. 通过热导性测试，结果显示为翡翠。
2. 通过偏光镜检测，显示该玉镯为多晶质宝石，符合翡翠的晶体特点。
3. 通过查尔斯滤色镜检测，无红光反应，说明未经过染色处理。
4. 通过鉴定专用的紫外线灯检测，在紫外线照射下该玉镯并未发光，显示其为A级天然翡翠。
5. 通过放大镜检测，该玉镯具备翡翠的结构和特征。
6. 通过硬度检测，该玉镯硬度为6.5，与翡翠硬度相符。
7. 通过普通光谱仪检测，吸收光谱所在区域代表其为天然A级翡翠。
8. 该玉镯的制作和打磨工艺均属上乘，明显有别于其他低质量玉镯。

如要最终检测该翡翠是否经过极其复杂的染色或处理，需要用到更加昂贵的仪器设备，如拉曼光谱仪和傅里叶变换红外光谱仪。

出于谨慎考虑，我们的专家建议根据检测结果将该玉镯作为真翡翠销售，但不担保其为A级翡翠，因为最终的保证需要使用更先进的设备进行进一步的检测。

总而言之，该玉镯为翡翠，并可能为A级翡翠，因为所有的检测显示其极可能是A级翡翠。

我们在升级我们的检测设备之后会提供一份A级翡翠证书，在那之前我们会保持谨慎。

我们发现有些商家在无任何有效检测的情况下就将商品标注为A级，我们也会谨防其他商家抄袭我们的信息。

我们所有的检测结果都有证据支撑，必要情况下可向相关人士出示照片证据。

玉龄：很可能属20世纪中后期。

来源：一位著名的中国传世珍品私藏家的藏品。

该商品属于一位澳大利亚人的家传藏品中的中国珠宝系列，我们之前已收到其部分藏品，并从上周日起上架了几件软玉藏品。

整套藏品以中国古玩为主，是在多年的游历和收集中积攒而成。

性状：状况良好，无损，未经修补。

该商品有天然内含物和变异及微小的天然裂纹。

尺寸：

外直径：69毫米（2.7英寸）

内直径：54毫米（2.1英寸）

宽度：15毫米（0.6英寸）

测量可能略有偏差。

销售信息

阿德雷德亚洲拍卖行决定出售的这个玉镯属于一个艺术品精选系列，该系列由澳大利亚各种私家藏品构成，包括玉件、铜件和其他源于中国的藏品。

我们最近所获取的这些藏品经由数代人精心收藏，多是收藏家在游历期间从世界各地收购而来。

该商品在网上出售，但是作为藏品所有人或继承人的代表，如有意外事件发生，我们保留撤售该商品的权利。如发生此类事件，我们会对重新销售该商品提供5%~10%的折扣。

玉石信息

玉石分为两类，软玉和硬玉（翡翠），以软玉更为常见。

软玉早在公元前5000年便出现在中国，而翡翠在中国最早的记录出现在18世纪中期大量翡翠原石进口到中国之后。翡翠略硬于软玉，多用作珠宝饰品。

翡翠比软玉更受欢迎，按重量计算价值更高。

上等原石，尤其是翡翠原石的价格从1980年开始飙涨。

白河玉（非翡翠）原石历年的市场价格为：

1980 年：每千克 100 元

2000 年：每千克 30 000 元

2010 年：每千克 300 000 元

一些更为稀少、更受追捧的原石在 2010 年的售价高达每千克 200 万元。

Part Ⅲ Extended Learning

◆ **Dictation**

1. reserves
2. with significant jade deposits
3. Under the circumstances
4. highly sophisticated
5. unearthed
6. jade tools and ornaments
7. similar craft of polishing stones
8. ultimate form of civilization
9. very hard texture
10. the key factor is its hardness

◆ **Read More（Translation）**

Passage 1

中国玉文化

许慎（约58—147 年）在中国的第一本词典《说文解字》中将"玉"定义为"美丽的石头"。玉一般分为软玉（和田玉）和硬玉（翡翠）。翡翠在清朝年间（1271—1368 年）从缅甸传入中国，在此之前中国只有软玉。传统上的玉特指软玉，故又称传统玉石。硬玉在中文中称为翡翠，在中国目前翡翠比软玉更受欢迎，价值更高。

玉的历史与中国的文明一样悠长。考古学家发现了以浙江省的河姆渡文化为代表的新石器时代早期（约公元前7000 年）的玉件，还有以西辽河流域的红山文化、黄河流域的龙山文化以及太湖地区的良渚文化为代表的新石器时代中后期的玉件。玉自古盛行至今。

中国人爱玉，不仅在于玉的美，更重要的是因为玉的文化、寓意和灵性。孔子（公元前551—479 年）曾说玉有十一德，曰：

"夫昔者，君子比德于玉焉。温润而泽，仁也；缜密以栗，知也；廉而不刿，义也；垂之如队，礼也；叩之，其声清越以长，其终诎然，乐也；瑕不掩瑜，瑜不掩瑕，忠也；孚尹旁达，信也；气如白虹，天也；精神见于山川，地也；圭璋特达，德也；天下莫不贵者，道也。诗云：言念君子，温其如玉。"

可见，玉在中国文化中的地位不一般。古言道："黄金有价而玉无价。"

玉代表美丽、优雅和纯洁，很多中国成语和习语用玉比喻美好的事物和人物，譬如玉洁冰清、亭亭玉立和玉女。中国人常用玉命名。

中国有很多与玉相关的故事，最著名的莫过于《和氏璧》和《完璧归赵》。璧即玉也。和氏璧讲的是一位叫卞和的人反复将一块璞玉荐赠君王而惨遭刑辱的故事，该璞玉最终被楚文王（约公元前689 年）确定为稀世之玉，并命名为和氏璧。完璧归赵是该玉的后续故事。战国时期（公元前475—221 年），当时的强国秦国的秦王想用15 座城池交换赵国的和氏璧，结果未能如愿，和氏璧被完好无损地归还赵国。玉在古代不仅价值连城，而且还是权力的象征。有意思的是，道教的至尊神明被称为玉皇大帝。

玉被制成祭器、工具、饰品、餐具等，一些古乐器也是用玉做成的，如玉笛、玉箫和玉钟。在古代，中国人觉得玉很神秘，因此玉器被普遍用作祭器，并常用于陪葬。为保护尸骨不朽，汉朝中山靖王刘胜（公元前113 年）死时着玉衣下葬，该玉衣由 2498 片玉片用金丝编缀而成。

玉在中国有丰富的文化底蕴，我们只是略知皮毛。总而言之，玉在中国文化中象征着美丽、高贵、圆满、坚贞、权力和永恒。

Passage 2

购买玉饰的五条建议

越来越多有天分的业余爱好者和纪念品寻觅者购买玉石饰品。然而,像以前在香港和台湾这样的地方不难淘到低价玉饰的时代早已改变。现在想用合理的价格买到玉饰已经越来越难,更别说低价了。

香港和台湾依然是玉器贸易中心,但目前市场上充斥着有时被作为真玉出售的仿冒玉器和劣质玉器,很难买到便宜的真玉。香港和台湾的玉在价格上较有优势,在这里买玉你可以少花钱,只是没有以往传说中那么划算了。归根结底,真玉是昂贵的。

到了香港,一个必去之地就是尖沙咀的玉器市场,那里卖家云集。摊档上展出的大部分是廉价的饰品和纪念品,但也有一些真正的商家在做玉器生意。

1. 做好功课

无论买什么,尤其是做大笔的投资,要先明确购买目标。玉器种类繁多,从不透明的黄色玉石到深绿玉石,五花八门,价格不一(深绿和纯白玉器标价最高,相对较少)。不管你想买哪种玉,一定要跟商家们确认价格,在你自己的国家或是任何一个卖玉的地方都是一样。要货比三家,这会让你对某些商品的市场价值有所了解,使你在讨价还价的时候心里有底。

2. 不敢相信这个价格?那就算了

虽然这是一条简单的建议,还是有很多人会被难住。如果有商家以远低于市场价的价格出售一件玉石,并宣称其为纯正的墨绿翡翠,或是有着千年的历史,那赶紧走开。没有这样的便宜货,但是确有这样的骗局。

3. 检验玉石

我们并非玉石专家,所以在做任何类型的大笔玉石投资之时,肯定需要请独立的专家进行检测。但是,还是有一些方法可以帮你识别真玉。买玉的时候,玉石摸起来须是光滑、冰凉的。真玉硬度高,因此表面不应有划痕。如果你能用指甲划出划痕,那就是假玉。假玉通常比真玉轻一些。

4. 索要证书

对于昂贵的玉件,商家需提供鉴定证书,说明玉件经过检测和验证。玉石等级分为 A 到 D 级,全球通用,但是大部分商家只提供 A 级玉石证书。值得注意的是,所有等级(包括 A 到 D)都是"真玉",只是质量和价值不同。B 级和 C 级玉石经过化学处理,并可能染过色。

5. 无须闻仿玉色变

玻璃、塑料和很多各类矿石都可以用来制作仿玉,价格可能还高于最低等的不透明玉石。只要商家坦诚真伪,你没必要拒绝仿玉。有些仿玉件非常漂亮。

Unit 8　Pearl

Starting Out

◆ **Match Words with Pictures**

1. oyster　2. round pearl　3. shell　4. baroque pearl　5. teardrop pearl　6. black pearl

◆ **Check Your Knowledge**

中文	中文
光泽	淡水珍珠
海水珍珠	天然珍珠
养殖珍珠	圆形珍珠
泪滴形珍珠	吊坠

续上表

中文		中文	
软件动物		珍珠贝	
珍珠层		评级，等级	

Part Ⅰ Communicative Activities
1. e 2. d 3. a 4. f 5. b 6. c

Part Ⅱ Read and Explore

◆ Check Your Understanding

Ⅰ.

Aspects		Description
luster		the sheen and iridescence of the pearls' surface and is determined by how the layers of nacre absorb and reflect light
color	light pearls	white, ivory with rose or silver overtones
color	dark pearls	black with green or blue overtones
shape		symmetrical pearls such as round, oval or tear drop ones
size		larger pearls are obviously harder to grow, making them more scarce and valuable

Ⅱ. 1. Pearl is formed within the body of an oyster or other small mollusk. This formation takes place when an irritant of some sort enters the mollusk's body or shell.

2. The author introduces two types: cultured pearls and natural pearls.

3. Luster is the sheen and iridescence of the pearls' surface and is determined by how the layers of nacre absorb and reflect light. Color refers to light colors and dark colors.

4. Grade refers to the judgment of pearls according to their qualities and appearances. Grade, based on their various standards, can vary quite differently.

5. Your budget will also greatly impact the pearls you choose. Cultured pearls are much more common than natural pearls, making them more affordable. If price is no object, you can choose the best natural pearls you can afford.

◆ Language Focus

Ⅰ. 1. budget 2. grade 3. layer 4. symmetrical 5. shell
6. nacre 7. sheen 8. luster 9. irritant 10. mollusk

Ⅱ. 1. take place 2. look for 3. determine 4. various 5. prior to
6. reflect 7. randomly 8. take into consideration 9. familiarize; with 10. be free of

Ⅲ. 1. in 改为 on 2. by 改为 on 3. in 改为 to 4. about 改为 for 5. 去掉 to
6. on 改为 at 7. with 改为 to 8. for 改为 of 9. for 改为 by
10. in holiday 改为 on holiday

Ⅳ. 1. 如果你打算购买珍珠饰品，在此之前需对珍珠有所了解。

2. 这个形成的过程从某种刺激物进入软体动物的身体或躯壳开始。

3. 不管这个刺激物是寄生虫、沙子或是其他异物，软体动物都会产生一种坚硬的、保护性的外层包裹住异物。

4. 光泽度是珍珠表面的光泽和结构色的展现，是由珍珠层吸收并反射光线所产生的。

5. There are two main types of pearls in the market: cultured pearls and natural pearls.
6. Pearls are graded on their quality and appearance.
7. To choose the best pearls, first examine their luster and ask about their grade.
8. Light colored pearls can also be found with rose or silver overtones, making them more desirable to many.

◆ **Text**（Translation）

如何选择优质珍珠

珠珠在珠宝饰品中较为流行，非常美丽。虽然大多数珠宝首饰都镶有珍珠，但是很少人知道如何选择优质的珍珠。如果你打算购买珍珠饰品，那么就应该在购买之前学习有关珍珠的来源和特点，这样才能选出优质的珍珠。

珍珠的来源

为了选择最优质的珍珠，我们必须了解珍珠。珍珠形成于牡蛎或其他小软体动物的体内。软体动物的身体或躯壳受到刺激后就会产生珍珠。无论这种刺激来自寄生虫、一粒沙子还是其他异物，受到刺激后的软体动物都会在异物外部形成坚硬的保护层。这种防护材料就是珍珠层，主要由碳酸钙组成。珍珠层会不断生长，一层一层，直到形成珍珠。珍珠通常形成于淡水和海水软体动物中。

珍珠的类型

珍珠主要分为两种。今天市场上所见的大部分珍珠都是人工养殖的珍珠。养殖珍珠由人工创造和培育。养殖户通过给软体动物植入刺激异物来促进珍珠的生长。软体动物受到刺激之后便开始分泌碳酸钙，形成珍珠层。另一类珍珠就是天然珍珠。它们是随机形成的，相比养殖珍珠而言较为少见。天然珍珠较为少见的原因在于，必须有来自自然界的刺激异物寄居在软体动物体内才能使它形成，不可人工干预。正因为天然珍珠较为罕见，所以其价值也高于人工养殖的珍珠。既然已经对珍珠有所了解，我们便可以开始学习如何挑选优质珍珠了。

选择优质珍珠

选择珍珠时，首先要看它们的光泽度。珍珠的光泽度是珍珠表面的光泽和结构色的展现。光泽度是由珍珠层吸收与反射阳光的方式所决定的。光泽和结构色越好往往代表珍珠的品质更出色。颜色是选择珍珠时应考虑的另一个因素。浅色珍珠通常分为白色和象牙色。浅色珍珠能够呈现出粉红色或银色，这种珍珠品质较好。此外也有黑珍珠，这种珍珠为暗含绿色或蓝色的黑色。黑珍珠非常罕见，所以市场价值更高。但实际上，珍珠也可以人工染成各种颜色。珍珠的形状也非常重要，优质珍珠形状对称，表面光滑，没有任何缺陷。形状对称的珍珠有多种形状，例如圆形、椭圆形或水滴形。珍珠的大小也决定了珍珠的价值。体积较大的珍珠形成过程艰难，更为稀缺，其市场价值也就更高。珍珠按照质量和外观可以分为不同的等级。珍珠的分级标准各不相同，最好向珠宝商询问有关珍珠的细节信息。

底线

选择珍珠时，要首先检查它们的光泽度，询问珍珠的等级。这些都是选择优质珍珠时所必须考虑的因素。然后，我们需要确保珍珠形状对称，外观一致。选购彩色珍珠时，必须查明珍珠的来源。此外，个人预算也是重要的影响因素。相比天然珍珠而言，人工养殖的珍珠价格低廉，处于多数人的承受范围之内。

如果资金充足，则可以选择优质的天然珍珠。

既然已经了解了有关珍珠的基本知识，我们就有能力挑选出优质的珍珠。在做出购买决定之前，要多看几家珠宝店，这样就可以有更多的选购余地。

Part Ⅲ Extended Learning

◆ **Dictation**

1. Round	2. luster	3. blister	4. baroque	5. nacre
6. cultured	7. Natural	8. drilled	9. oyster	10. millimeter

◆ **Read More（Translation）**

Passage 1

珍珠选购新手指南

珍珠是一种有机宝石，形成于生物体内。每一颗珍珠最初都是一块沙砾或者一粒沙子进入到海洋或淡水贝类体内，例如牡蛎和蛤蜊。贝类受到刺激后启动防御机制，形成珍珠层，也称为珍珠母，然后逐渐变厚，最终形成珍珠。

我们可以根据珍珠的来源和形状对其进行分类。

按照来源对珍珠进行分类

天然珍珠

入侵生物进入软体动物的外壳内部之后，软体动物便会受到刺激，围绕入侵生物形成与洋葱皮相似的保护层，最终形成珍珠。天然珍珠的形状主要取决于入侵生物的形状。

天然珍珠较为罕见，价格相当昂贵。珠宝商通常以克拉为重量衡量其市场价值。目前市场上最为常见的天然珍珠就是复古珍珠。

人工培育珍珠

与天然珍珠相似，人工养殖的珍珠也生长于软体动物体内，但却在人工干预下生长成型。养殖户打开动物外壳，然后植入一个物体。植入物体的形状将最终决定珍珠的形状。

随着时间的推移，人工植入的物体被珍珠层覆盖。珍珠层的厚度取决于软体动物的类型，软体动物所在水域，以及养殖户收取珍珠之前植入物在软体动物体内的滞留时间。珍珠层越厚，珍珠的质量和耐久性就越好。人工培育的珍珠通常以毫米为单位出售。

海水珍珠

海水珍珠主要来源于海水中的软体动物。海水珍珠也分为天然珍珠和人工培育珍珠。

淡水珍珠

淡水珍珠主要来源于淡水中的软体动物，也就是生活在河流或湖泊中的软体动物。

按照珍珠形状对其进行分类

球形珍珠是圆形的。从传统意义上来说，这是最为理想的珍珠形状。珍珠形状越圆市场价格越高。

对称珍珠包括梨形珍珠和其他对称形状的珍珠，它们左右对称，但却不是圆形。

巴洛克珍珠是指形状不规则的珍珠。虽然价格最为便宜，但是形状独特，非常美丽。

Passage 2

如何清洗珍珠

大多数人佩戴的珍珠都是人工培育形成的。养殖户在牡蛎或蛤蜊体内植入珠子或其他物体，牡蛎或蛤蜊在该物体外部生成珍珠层，这层光泽形成珍珠独特的外观。

珍珠保养

人工培育的珍珠往往具有较厚的珍珠层，但是这种珍珠相比宝石而言仍然较为脆弱，所以需要细心保养，以保证珍珠的光泽。

涂抹化妆品和喷洒香水之后再佩戴珍珠饰品，以保持珍珠饰品的清洁。

涂抹护手霜和润肤霜之前务必把珍珠戒指摘下来。

佩戴后即使用柔软无绒布擦拭珍珠，然后收起珍珠。可以使用湿润或干燥的无绒布擦拭珍珠。如果珍珠较为潮湿，则应晾干后再行存放。

可以使用温和的肥皂水溶液清洗珍珠。

不得使用氨溶液或强力洗涤剂清洗珍珠。

不得使用超声波清洗机清洗珍珠饰品。

不得使用研磨性清洁剂或砂布清洗珍珠，否则会磨损珍珠层，损坏珍珠质地。

珍珠存放事宜

不得把珍珠与其他珠宝饰品共同存放，否则取出珍珠饰品时可能会被金属或宝石饰品刮花。

使用带有珍珠专用插槽的首饰盒，或者把珍珠放在麂皮和软袋中。

应定期重新串连珍珠项链，确保串连珍珠的丝绸或尼龙线完整无损。

Unit 9　Gemstone Design

Starting Out

◆ **Match Words with Pictures**

1．blue zircon　2．twinkle oval amethyst　3．natural top green emerald oval

4．blue sapphire octagon shape　5．round brilliant zircon　6．flower round cut ruby

◆ **Check Your Knowledge**

1．table　2．star facets　3．upper-girdle facets　4．lower-girdle facets　5．pavilion main facets　6．culet

7．girdle　8．crown

Part Ⅰ　Communicative Activities

1．d　2．a　3．e　4．b　5．c

Part Ⅱ　Read and Explore

◆ **Check Your Understanding**

Ⅰ．1．handmade；die stamping and jig assembly manufacture

2．cut；stamped

3．creativity；art

4．earrings；pendants；brooches

5．replicate natural objects；create interesting shapes from manmade

Ⅱ．1．Good design is especially crucial and significant since any design mistakes are very costly, both in terms of labor and tooling costs.

2．The raw material is first made up to the design using ready formed metals, then mounters fashion and assemble the pieces of jewelry, and then pass on to the finisher and setter.

3．Only where high numbers of the same item are being produced.

4．Both are cost-effective in mass production.

5．Electroforming is very effective when requirements call for extreme tolerances, complexity or light weight.

◆ **Language Focus**

Ⅰ．1．gravel　　　　2．lightweight　　　3．fabricate　　　4．snag　　　　5．increasingly

6．inherent　　　7．indistinguishable　8．crucial　　　　9．replica　　　10．synthesize

Ⅱ．1．subsequently　2．resurge　　　　　3．a strip of　　　4．filing　　　　5．scrap

6．voluminous　　7．raw　　　　　　8．sterling　　　　9．Replica　　　10．patterned

Ⅲ．1．删去 of　　　　2．删去 if　　　　　3．for 改为 in　　4．删去 since　　5．删去 long as

6．删去 in　　　　7．删去 has　　　　8．Thanks to 改为 Because of/Owing to/Due to

9．Seeing 改为 Seeing that　　　　　　10．删去 that

Ⅳ．1．大多数好的设计师都是熟练的珠宝工匠，他们了解原材料以及其自身的局限性，也明白功能性问题。

2．当一个部件完成后，不管用哪种方法，这个部件被传至整理工那里进行打磨抛光，最后，如有必要，会交给镶嵌工来完成珠宝镶嵌。

3．对大多数人来说，珠宝设计让人想到创新和艺术，在艺术家的工作坊里与物理和技术相关的概念似乎没有立足之处。

4. 电铸技术主要的弊端在于电镀技术要求专门用于电镀法的特定设备。
5. A good cut showcases the gemstone's color, diminishes its inclusions, and exhibits good overall symmetry and proportion.
6. Most gems that are opaque rather than transparent are cut as cabochons rather than faceted, including opal and moonstone.
7. Electroforming allows for the realization of designs that are not possible by other techniques, at affordable prices.
8. Jewelers do not restrict themselves to one particular method, and, depending on what is to be made, will use, for maximum efficiency, any of the three methods, in combination, to achieve desired quality within a price range.

◆ **Text（Translation）**

珠宝的生产过程和技术

自古以来，熟练的金匠利用技艺和简单的手工工具制造珠宝，即使在今天，世界各地依然有不少珠宝是在普通工作室用手工的金属加工技术和有限的机械制成的。但是，新的大批量生产方式正在越来越多地被采用，包含CAD（电脑辅助设计）的精准工程技术和能生产高质量的、与传统的手工制作的珠宝没有区别的激光技术。任何装备都应围绕使用现代设备的熟练工匠为核心而生产。

好的设计是基础，更是大批量珠宝生产的关键，因为无论从劳动力成本还是工具成本上讲，设计错误的代价很大。大部分好的设计师都是熟练的珠宝工匠，他们了解原材料和其自身的局限性，也明白功能性问题。比如，项链必须垂落得刚好，戴起来舒服，耳环不能太重，手镯不能摩擦到肉，戒指不能勾衣物等等。一个好的设计师必须注意到潮流趋势，能够预测出两年后流行的趋势。

有三种生产珠宝的方法：手工、模具冲压和夹具装配制造或铸造全部或部分。当一个部件完成后，不管用哪种方法，这个部件被传至整理工那里进行打磨抛光，最后，如有必要，会交给镶嵌工来完成珠宝镶嵌。

珠宝匠人们并不局限于某一种方法，而是根据要生产的珠宝来使用这三种方法中的任何一种或几种以便最高效地在一定价格区间内获得想要的质量。

手工珠宝

安装工首先根据设计用已经成型的金属来生产出片状的、线状的、管状的原材料和零件（现成的联结部件、搭扣、底座），所用的工具都是传统的工匠工具：锤子、钻、钻孔机、沙砾、锉刀和热源，现在还有喷灯。安装工借助这些工具使用并组装这些珠宝部件，然后再把这些交给精整工和镶嵌工。

机制珠宝

机械生产的基本方法就是切割零部件并用单独的钢模盖印，这些单独的钢模是为每一件珠宝的每一个零部件而生产的。由于模具制造很昂贵，只有大批量生产同样产品时才划算。一旦零部件生产出来，它们被装在夹具上等待最后的装配过程，而最后的装配过程半熟练的安装工也可以完成。在大批量生产中会使用复杂的连续模具。在这一过程中，一列半成品通过单个的模具被生产出来，并以最终形状成品。

铸造

"失蜡"铸造法（用原件的蜡复制品，也被称为熔模铸造法）自公元前4000年就已开始使用，使用这种方法很容易大批量生产，优点在于资金成本低廉。有时，为了制作一件设计复杂或精巧的珠宝，必须要焊接许多铸造件配件。焊料必须是金、银或铂合金，且必须符合鉴定要求。铸造法正在重新流行起来。

电铸法

对大多数人来说，珠宝设计让人联想起创新和艺术。在艺术家的工作坊里与物理和技术相关的概念似乎没有立足之地。但电铸法这一新技术为全世界的设计师们打开了机会之门。任何认真的珠宝设计师都值得花费时间来了解这一新技术背后的过程。

什么是电铸法？电铸法是高度专业的金属部件加工过程，它基于原型（称为心轴、模型或图案）使用电镀槽里的电沉积来进行加工，原型随后被取出。从技术上讲，泡入了金属或金属物的电解质溶液的金属中是有电沉积的，通过控制电沉积来合成金属物件的过程就是电铸法的过程。更简单一点讲，就是在金属表面或任何表面通过涂上带有金属分子的涂料而使之能导电的表面镀上了一层金属外衣。本质上就是从镀层上制作了一个金属部件。

为什么要使用电镀法？和其他基本的金属制作过程（铸造、锻造、冲压、拉深、加工和制造）相比，如果有加工耐受性极强，结构极复杂和重量极轻的产品需求，电镀法非常有效。制作逼真的导电图案衬底所固有的精度和分辨率能做出更好的、公差更小的几何形状，同时维持超高的边缘轮廓精度和抛光。电铸法制成的金属极纯，再加上它精细的晶体结构，它比锻制金属拥有更优的性能。

电镀法有无尽的优势。珠宝可以被加工成24K金，也可以被加工成8K至18K。电镀过程使我们可以做出形状复杂的、三维的、薄且中空的珠宝。我们可以依照各种设计制造产品，包括大而轻的、壁厚均匀的现代造型，而且在制作过程中不损失金属，也不产生废料。其主要的弊端在于电镀技术要求专门用于电镀法的特定设备。

电镀技术无疑会对珠宝行业产生影响。对于薄的、中空的、轻的、用料繁多的复杂的三维造型，电镀法无与伦比。电镀法能以可承受的价格生产出其他方法无法生产的设计。在它的许多应用中，电镀法可以复制大自然的物体，比如，树叶、花、贝壳、果实，也能从人造物体中创造出有趣的造型。应用电镀法制成的经典珠宝有耳环、吊坠、手镯、链条、项链、小装饰品、扣环和手链。

Part Ⅲ   Extended Learning
◆ Dictation

1. collections
2. showcasing
3. an outfit
4. diamond headpieces
5. reflection
6. intricately
7. rows of
8. brand ambassador
9. a queen necklace
10. craftsmanship

◆ Read More（Translation）

Passage 1

宝石加工

尽管可以找到一些好的晶体，但大多数宝石都经历了水蚀作用损坏或呈鹅卵石形。因为宝石的光学性质可以产生最佳效果，所以适当的刻面型切工能使宝石具有更高的价值，也更加漂亮。宝石的加工旨在修正晶体使之匀称。

宝石的切割质量可以对它的外观产生重大影响，而对克拉价格影响很小。珠宝商和精明的珠宝消费者都会为珠宝的美丽付出代价因而特别注重宝石加工。

你能怎么分辨宝石加工的好坏呢？训练眼力的最简单方法是先看差的加工再看好的加工。上图的三颗红宝石，颜色和净度都很好。中间那颗红宝石看起来更好，因为所有射入宝石的光线都被肉眼观察到。看看颜色是如何均匀地分布于宝石表面的。这颗红宝石更有生机也很闪耀，如同光线在刻面上跳跃。相反，另外两块宝石有暗色区域，光线没有沿适当的角度全部反射回观察者的眼中。

加工款式也可以增加宝石的美丽程度。看看上图的这两块宝石：它们的大小、形状、品质和颜色都相同，而外形却显示出明显的不同，因其中一个是标准的祖母绿切工，另一个是重子切工，背面有更多的小刻面。

除了这些标准的宝石切割形状外，珠宝设计师们正在发明针对个性化款式的一些新的宝石切工方法，如一些具有不寻常几何造型的刻面宝石、雕刻宝石、刻面与雕刻相结合的宝石。

一个好的切工能够展示宝石的颜色、减少夹杂物并呈现良好的整体对称性及切工比例。由于宝石的颜色会变，因此，当谈到最好的火彩或颜色时没有硬性的几何学标准。宝石，尤其是稀有宝石，有时其切割只关注尺寸而不考虑颜色因素。例如,刚玉的变种宝石蓝宝石和红宝石,加工时会尽可能保重优先于切割漂亮程度,此时可能会出现色带或生长纹。

对一颗颜色较深的宝石,其最好的切工可能是比一般宝石切得浅一些,允许更多的光线穿透到宝石中;而对颜色较淡的宝石,切得厚一些可能对颜色更有利。

查看镶嵌的宝石,确保所有刻面都是对称的。一个不对称切工的冠部代表宝石的切工质量低。在所有情况下,切工好的宝石都是对称的,反射光均匀地穿过表面,且抛光光滑,没有任何刻痕或擦痕。像钻石,颜色级别高的宝石通常都有台面、冠部、腰棱、亭部和钻尖。变彩欧泊是个特例,绝大部分为圆弧形切工。

大多数不透明的宝石,包括欧泊、绿松石、缟玛瑙、月光石等,一般都加工成弧面型,而透明宝石一般加工成刻面型。你也会看到品级较低的蓝宝石、红宝石和石榴石加工成弧面型。如果宝石材料的颜色非常好但不够透明或洁净以加工成刻面宝石,那么也可将其加工和抛光成非常吸引人的弧面型宝石。因莫氏硬度小于7的宝石容易被粉尘和研磨砂里的石英划伤,较软的宝石材料也通常加工成弧面型,弧面型宝石上显示的微细划痕比刻面型宝石上的要少得多。

Passage 2

判断镶座的质量

既然已经找到了理想的宝石,那么你所需要做的就是确保它能完好展示又安全地安放。

要判断珠宝镶座的质量,需要密切注意细节。放珠宝的金属镶座是否抛光平整光滑,不会勾住衣物?珠宝在镶座里是否牢固?镶座是否抛光很好,没有小毛口或凹痕?

便宜的珠宝往往比较轻,给你物超所值的视觉体验。如果一件珠宝很轻,要特别注意:放珠宝的金属爪是否牢固?金属爪夹珠宝紧不紧?如果因此而丢失了珠宝,你一定不会因省下制作金镶座的钱而感到开心。

如果珠宝是黄金,那么有没有开数标记?公司商标是否贴在它旁边?若是如此,那就表明公司向你保证开数和标记的是一致的。

买项链时,要确保它长度合适。可以试戴或让店员展示,这样你可以检查它和肤色是否搭配。对于耳环,要确保它们比较合适,不要向前倾斜。对称的设计应该左右互为照应。

购买珠宝时的最后一点提示:如果一件珠宝做工精良,背面抛光也会很好。

如果你买了一件礼物而自己又不确定这件珠宝的样式,干吗不直接把这件完美的珠宝放进一个黑色的天鹅绒袋子,让幸运的收礼人来设计她自己的完美镶座;许多珠宝商提供定制服务。宝石比言语更有用。你可以选择一件能表达出你想借此礼物所要传达的意思的珠宝。

Unit 10   Jewelry Commerce

Starting Out

◆ **Match Words with Pictures**

1. tourmaline   2. Cat's eye   3. prehnite   4. moonstone   5. topaz
6. obsidian   7. rhodochrosite   8. garnet   9. sugilite

◆ **Check Your Knowledge**

1. When experts appraise something, they decide how much money it is worth.
2. If you estimate a quantity or value, you make an approximate judgment or calculation of it.
3. If you evaluate something or someone, you consider them in order to make a judgment about them, for example about how good or bad they are.
4. A gem certificate comprises identification data based upon the examination of precious stones and precious metals used in jewelry.
5. If you describe something as invaluable, you mean that it is extremely useful.

6. If you describe something as valueless, you mean that it is not at all useful.
7. If you appreciate something, for example, a piece of music or good food, you like it because you recognize its good qualities.
8. The brand name of a product is the name the manufacturer gives it and under which it is sold.
9. A trademark is a name or symbol that a company uses on its products and that cannot legally be used by another company.
10. If something has a price tag of a particular amount, that is the amount you must pay in order to buy it.

Part Ⅰ Communicative Activities

1. e 2. d 3. a 4. c 5. b

Part Ⅱ Read and Explore

◆ **Check Your Understanding**

Ⅰ. 1. gem variety; color; clarity; cut and polish; size; shape; treatment; origin; fashion; supply chain
   2. color
   3. proper proportions
   4. heating; fracture-filling; radiation; diffusion

Ⅱ. 1. Some gemstone varieties command a premium price in the market, due to their superior gemstone characteristics and rarity. Other varieties are abundant in many locations around the world, and prices are much lower.
   2. In colored gemstones, color is the single most important determinant of value. Ideal colors vary by gem variety of course, but generally the colors that are most highly regarded are intense, vivid and pure.
   3. A gemstone that is perfectly clean, with no visible inclusions, will be priced higher. In general, the cleaner the stone, the better it's brilliance.
   4. Some gems, such as blue sapphire, are always in fashion. But some gems become fashionable for short periods, with resulting price increases.

◆ **Language Focus**

Ⅰ. 1. negative   2. obscure   3. translucent   4. flimsy   5. salient
   6. emerald   7. determinate   8. polish   9. diffusion   10. spotlight

Ⅱ. 1. vary   2. literally   3. influenced   4. concise   5. due to
   6. set a price for 7. have an effect on 8. in general 9. interfere with 10. tend to

Ⅲ. 1. got 改为 gets   2. become 改为 becomes
    3. will work 改为 work   4. practised 改为 practise
    5. felt 改为 feels   6. 将第二个 there is 改为 there is in it
    7. will flatter 改为 flatters 8. has worked 改为 worked
    9. more happy 改为 happier   10. you've 改为 you'll

Ⅳ. 1. 不同的宝石每克拉的价格可能差距非常大,理论上 1 克拉宝石的价格变动范围从 1 美元到数万美元。影响每克拉宝石价格的因素有许多。
    2. 太深或太浅颜色的宝石在价格上会比中间色调的那类宝石便宜。所以矢车菊蓝的蓝宝石会比墨水蓝或浅蓝的蓝宝石的价值高得多。
    3. 诚然,宝石净度级别越高,其价值就越高,但有些不影响宝石火彩和闪耀的内含物不太会影响宝石的价值。
    4. 严格来讲,不论产自哪个国家和哪个地区,一块好的天然宝石就是一块好宝石。但事实上某些地区如缅甸、克什米尔、斯里兰卡、巴西出产的宝石在市场上的价格会高一些。
    5. Gemstones can pass through many hands on the way from the mine to the consumer, and the more brokers

and distributors that handle the product, the higher the final price will be.
6. Rounds are much less common than ovals, since ovals are usually cut to preserve as much weight of the raw material as possible.
7. A number of popular gems, such as tourmaline, spinel, amethyst and garnet are almost never treated.
8. Some very fine gems, such as natural spinel, actually have lower than expected prices because limited supply means that the gems are not promoted heavily in the market.

◆ **Text**（**Translation**）

宝石的价格

不同的宝石每克拉的价格差距可能非常大，理论上1克拉宝石的价格变动范围可从1美元到数万美元。影响每克拉宝石价格的因素有许多，这里简要总结10个影响宝石价格的因素。

宝石品种

一些宝石品种，如蓝宝石、红宝石、祖母绿、石榴石、坦桑石、尖晶石和变石，由于它们的优质宝石特性和稀缺性，在市场上售价颇高。其他种类，比如各种石英，在世界多个地区大量存在，其价格就很低。宝石的种类大致决定了其价格范围，每块宝石的特性在很大程度上也影响每克拉的价格。

颜色

对有色宝石而言，颜色是唯一影响价格最重要的因素。理想的颜色因宝石的种类而异，但大体上颜色浓重、鲜艳、纯净是最受推崇的。太深或太浅颜色的宝石在价格上会比中间色调的那类宝石便宜。所以矢车菊蓝的蓝宝石会比墨水蓝或浅蓝的蓝宝石的价值高得多。

净度

一块非常干净、没有内含物的宝石价格相对更高。一般来说，宝石越干净，就越光芒夺目。诚然，宝石净度级别越高，其价值就越高，但有些不影响宝石火彩和闪耀的内含物不太会影响宝石的价值。同时要知道，有些宝石，如祖母绿，总是会含有内含物的。

切割和抛光

宝石应该按正确比例切割，以最大限度地将射入宝石的光反射到肉眼。但宝石工匠或珠宝商在切割特定宝石时，常常不得不做些妥协。如果宝石的颜色太浅，切得深一些可以获得更丰满的颜色。相反，将深色宝石切得浅一些可淡化成品宝石的颜色。但无论哪种情况，宝石的刻面应该整洁，表面应该被很好地抛光，没有划痕。

大小

对于某些宝石品种，比如石英类宝石，不管其重量多少，每克拉的单价相对稳定。但对于稀有宝石品种，价格并不随着重量的增加而线性增长。实际上如某些宝石，比如钻石，随着宝石克拉数的增加，价格呈现幂指数增长（暴涨）。按这个模式，一颗1克拉的宝石其价格可能是1000美元，而2克拉的就需要4000美元。尽管这个模式并不那么确切，但优质的红宝石、蓝宝石越大，就越具有每克拉价格更高的趋势。

不仅越大的宝石其价值越高，而且切割成珠宝交易中所知标准尺寸的宝石也会更贵。因为为了达到那个标准，更多的部分会被切掉。

啄形

某些啄形宝石会比其他啄形宝石的价格高一些，部分是因为需求，部分是因为切割成特定啄形时材料方面的问题。市场上一般圆形成品宝石的价格会高一些。圆形宝石较椭圆形不常见，因为椭圆形切工的宝石可以尽量保留原材料的重量。将宝石切成圆形会损耗更多的原石，对贵宝石比如红宝石、蓝宝石、变石等来说，会严重影响价格。

处理

宝石的处理如加热、裂隙充填、辐照和扩散会极大地改善很多宝石的外观。现在这些处理已经成了商业级宝石的例行工序。处理过的宝石总是比类似的未处理宝石便宜。但按惯例大多数宝石都会被处理，如红宝石和蓝宝石都经过处理，未经过处理的很少见，其市场价格也超出大多数顾客的承受范围。如果你非要买未处理过的宝石，也还是有很多选择的。许多流行的宝石如碧玺、尖晶石、紫晶和石榴石几乎都未经处理。

产地

严格来讲，不论产自哪个国家和哪个地区，一块好的天然宝石就是一块好宝石。但事实上某些地区如缅甸、克什米尔、斯里兰卡、巴西出产的宝石在市场上的价格会高一些。很难说这种价格附加是否合理，尤其是有很多好宝石来自非洲。

时尚

一些宝石比如蓝宝石，总是很流行。但有些宝石会在短期内流行，使得价格上涨。近来我们注意到，中长拉长石、硬水铝石就属于这种情况，也有很多人对金红石发晶很感兴趣。一些非常好的宝石，比如天然尖晶石，其价格实际上低于预期价格，因为有限的供给使得这些宝石在市场中没有被大规模推广。

供给链

宝石贸易是买卖交易，从宝石开采到零售，每个人都处在供给链中且都设法获利。宝石从开采者手中到最终消费者手中要经过很多次交易，经过的代理和分销商越多，其价格就越高。所以从不同的地方购买相同的宝石，价格可能相差 200% 之多。

Part Ⅲ　Extended Learning

◆ **Dictation**

1. ended on
2. domestic and overseas
3. jewelry hub
4. domestic trade fairs
5. by half
6. account for
7. in a long recession
8. whole economy
9. 200 billion
10. a host of collections

◆ **Read More（Translation）**

Passage 1

生辰宝石

石榴石

不同种类的石榴石有很多不同的颜色，包括红、橙、黄、绿、蓝、紫、棕、黑、粉红及透明。

其中最罕见的蓝石榴石，于 20 世纪 90 年代末期在马达加斯加贝基利首先被发现。此外在美国部分地区、俄罗斯及土耳其亦有其踪影。因为高钒含量的关系，它的颜色在白热光下会由日光下的蓝绿色转为紫色。

紫晶

紫晶是一种紫色的石英（二氧化硅），它的紫罗兰色具有独特的光彩，铁和其他微量元素是它的致色原因，也导致它独特的晶格结构。紫晶的硬度和石英一样，所以它很适合做珠宝。

纯净的石英，习惯上叫作水晶（有时也叫作无色的石英），它是无色的、透明的或者半透明的。常见的石英有很多种类的颜色，包括黄晶、玫瑰晶（粉晶）、绿晶、发晶、紫晶、烟晶和乳水晶。

海蓝宝石

海蓝宝石（来自拉丁语：aqua marina, 意为"海洋的水"）是一种蓝色或蓝绿色的绿柱石。在大多数出产普通绿柱石的地方都有海蓝宝石。

常见的绿柱石包括海蓝宝石、祖母绿、白柱石（也叫作"无色绿柱石"）、铯绿柱石（也称为"粉红绿柱石""玫瑰绿柱石"或"粉红祖母绿"）、红色绿柱石（也称为"红色祖母绿"或"朱红色祖母绿"）、金色绿柱石。

钻石

现今最常见的钻石用途是作为宝石用于装饰——此用途可以追溯到古代。在 20 世纪，钻石的四个特性"4C"作为厘定钻石价值高低的标准，即克拉重量、颜色、净度和切工。

翡翠
翡翠是一种罕见而有价值的宝石，同样，它的高价值间接推动了合成翡翠的发展。

大多数翡翠内嵌严重，因此硬度通常被归为脆弱型。

现今生产翡翠较有名的产地，首推哥伦比亚的三大产地。

珍珠
淡水珍珠和海水珍珠有时看上去很相似，但它们来源不同。

淡水珍珠是在不同种类的珠蚌里成长的，它的母体是生活在湖泊、河流、池塘和其他淡水中的无脊椎珠蚌。然而，大多数现售的淡水养殖珍珠来自中国。

海水珍珠是在不同的珍珠贝里成长的，它的母体是生活在海洋中的无脊椎珍珠贝。海水珍珠通常养殖在受保护的环礁湖或火山环礁中。

红宝石
红宝石是一种粉红色到血红色的宝石，是刚玉的一种，主要成分是氧化铝（Al_2O_3），红色来自铬（Cr）。红宝石其名来自拉丁语的"红"（ruber）。自然界中其他颜色的刚玉被认为是蓝宝石。红宝石被认为是四大名贵宝石之一，其他的是蓝宝石、翡翠、钻石。

红宝石的价格主要取决于红宝石的颜色。最亮、最宝贵的"红色"称为鸽子血红，其价值与其他颜色的红宝石有很大的差距。

红宝石在日光灯（顶部）照耀下和在绿激光灯（底部）照耀下呈现出不同的情况。红宝石会产生红色荧光。

橄榄石
橄榄石是为数不多的只有一种颜色的宝石，为橄榄绿色。橄榄石的绿色的强度和色彩取决于包含在水晶结构中铁的含量，所以橄榄石宝石的颜色从黄绿色到橄榄色再到棕绿色略有不同。其中最有价值的橄榄石是深橄榄绿色。

蓝宝石
蓝宝石（来源于希腊语的"蓝色石头"）是刚玉的一种，含有氧化铝（Al_2O_3），也含有其他微量元素，如铁（Fe）、钛（Ti）或铬（Cr）能为刚玉带来蓝色、黄色、粉色、紫色、橙色或微绿色。刚玉含有杂质铬（Cr），产生一种红色的色彩，因此而产生的宝石被称为红宝石。

星光蓝宝石是蓝宝石的一种，具有星光效应；具有星光效应的红宝石叫作"星光红宝石"。

欧泊
欧泊（蛋白石）是澳大利亚人的国石，全球欧泊供产的97%来自澳大利亚。

欧泊的内部结构使它可以衍射光彩，取决于形成条件，欧泊可以显现很多颜色。欧泊的颜色范围广，从无色到白色、灰色、红色、橙色、黄色、绿色、蓝色、洋红色、玫红色、粉红色、石蓝色、橄榄色、棕色和黑色。这些色彩中，红色与黑色是最罕见的，而白色和绿色是最常见的。

火焰女王欧泊也许是最著名的欧泊。

托帕石
纯的托帕石是无色透明的，但常因其中的杂质导致其不透明。典型的托帕石是葡萄酒色、黄色、淡灰色、橙红色或蓝棕色的。它也可以做成白色、淡绿色、蓝色、黄金色、粉色（罕见）、红黄色或由不透明做成透明。

帝国托帕石是黄色、粉色（天然的较罕见）、粉橘色的。

蓝色的托帕石是美国德州的宝石。天然蓝托帕石是相当少见的。

神秘的托帕石是无色黄玉，经过人工涂层，获得预期的彩虹效果。

绿松石

这种物质已被大家所公认的名称有很多，但是绿松石这个词，来源于16世纪的一个古老的法语单词"土耳其"。

这是一种罕见的具有高价值的较优等级的宝石，因其独特的色彩被作为装饰性宝石有几千年的历史。

Passage 2

GIA 和彩色钻石鉴定

关于 GIA

GIA（美国宝石学会）是极具规模的全球知名非营利宝石学教育及研究机构。

GIA 旨在帮助人们了解宝石知识。作为历史悠久的学术机构，GIA 不仅提供专业的钻石知识，更提供钻石鉴定证书，让消费者放心购买。

GIA 致力于保护钻石消费者，创立了 4C 标准和国际钻石分级系统™，并成为全球通用的钻石鉴定标准。作为标准的创立者，GIA 对宝石研究持续投入，在业界的权威性无出其右。

全球最负声望的珠宝零售商、博物馆、拍卖行和私人收藏家都信赖 GIA 等级鉴定师的专业知识，让他们为宝石进行评估、鉴定与核查。他们了解获取完整、公正和科学的宝石鉴定信息至关重要，也十分信赖 GIA 提供的鉴定。

GIA 彩色宝石鉴定

虽然绝大多数的钻石的颜色等级都在 D（无色）至 Z（淡黄或淡褐色）之内，但偶尔也有天然钻石呈现蓝色、棕色、粉红色、深黄色甚至是绿色。鉴于形成彩色钻石的地质条件极为特殊，所以彩色钻石十分稀有，更显珍贵。

与无色和接近无色的钻石不同，对彩色钻石进行亮泽度或火彩的鉴定较少，多数是针对色彩浓度进行的评估。颜色越深，越鲜明，价值越高，反之，价值越低。

GIA 通过色彩、色调和饱和度来描述钻石的颜色。色彩是指钻石的特征颜色；色调是指颜色的相对亮度或暗度；而饱和度是指颜色的深度和强度。彩色钻石等级鉴定师会在高度可控的观测条件下进行颜色对比，从 27 个色彩中选择一个，然后采用专业术语来描述色调和饱和度，诸如"淡彩""浓彩"和"艳彩"等。GIA 开发的颜色体系已在全球广泛使用。

彩色宝石鉴定

GIA 不仅因其专业权威的钻石等级鉴定闻名，还提供各类彩色宝石的鉴定服务。经过几十年的努力，GIA 建立了包含 10 万多个彩色宝石信息的数据库。使用该数据库和精密的分析仪器，GIA 的鉴定师和研究人员可以精确地确定宝石的种类及某些彩色宝石的原产地。此外，还会分析出宝石是否为合成的、仿品或是经过各式人工处理的。确定宝石颜色成因的一项非常重要的程序是确定颜色是天然的，还是经过加工处理的。

GIA 提供两种针对彩色钻石的鉴定证书。GIA 彩色钻石鉴定证书，其中所包括的 4C 标准全面信息与 GIA 钻石鉴定证书中的相同，但 GIA 彩色钻石颜色成因鉴定证书（也称颜色成因证书）仅限于颜色等级鉴定及颜色成因（天然或人工处理）信息。

珍珠鉴定与分级

一百多年来，珍珠养殖技术的发展为市场带来变革，人工养殖珍珠已基本上取代了天然珍珠。

诸如寄生虫等刺激物进入某种牡蛎、贻贝和蛤的体内，便会生成天然珍珠。作为一种防御措施，这些贝类软体动物会分泌我们俗称为珠母层的液体，来包裹刺激物。随着时间的推移，珠母层会形成天然珍珠。人工养殖珍珠的过程，是通过外界植入珠子或贝壳小粒来形成珍珠。

珍珠的天然形成数量较少，而通过人工养殖可以定期生成，从而使得珍珠为更多消费者认可、购买。但这也引发对于珍珠品质和鉴定方式的困惑。人工养殖珍珠的颜色、形状、大小差异繁多，等级鉴定也相应复杂。

与制定钻石等级鉴定标准一样，GIA 也创立了珍珠等级鉴定的标准和术语。GIA 的珍珠等级鉴定体系于 1998 年推出，该标准规定了如下 7 个珍珠价值因素™：大小、形状、颜色、光泽、表面品质、珠母层品质和配对。

References

［1］陈令霞，汤紫薇. 珠宝首饰专业英语［M］. 北京：化学工业出版社，2014.
［2］肖启云. 珠宝专业英语［M］. 武汉：中国地质大学出版社，2011.